Contents

How this book will help you

Exam practice – how to answer questions better

This book will help you improve your performance in your GCSE Intermediate or Higher Mathematics exam.

I mark hundreds of exam papers and find that students make the same mistakes on the same kinds of questions each year. For this reason, lots of students fail to get the grade that they are capable of achieving.

This book has been written to help you to tackle exam questions correctly so that you can achieve a really high grade. Students gain high GCSE grades in Mathematics through a combination of **good knowledge, good understanding and good examination technique.** It is the last of these with which this book will particularly help you.

Each chapter in the book is broken down into four separate elements, aimed at giving you as much **guidance** and **practice** as possible:

1 Exam Questions and Answers

The questions and answers at the start of each chapter are typical of the topic area. **In the 'How to score full marks' section, I show you how to tackle the questions in order to arrive at the correct answers.** This means that when you see questions like this on your exam paper, you will know exactly how to go about answering them in order to score full marks.

2 'Key points to remember'

This section identifies the **key elements** of the topic and the **essential knowledge** you need in order to answer exam questions. I have picked out what I think to be the most important points – but you will also need to use your own notes and textbook.

3 'Don't make these mistakes'

This section highlights common errors and misconceptions. Many of these errors seem quite logical when you are not actually sitting there taking the exam! It is a good idea, when you are into your last minute revision, to read quickly through all of these sections and make doubly sure you avoid these mistakes.

4 Questions to try, Answers and 'How to solve these questions'

Each chapter ends with a set of exam questions which I have carefully chosen as typical of those which you will meet in your exam. Don't cheat. Sit down and try to answer the questions as if you were in the exam. Try to remember all that you've read earlier in the chapter and put it into practice.

The answers are at the back of the book, and they are ones that would gain full marks. **Alongside each answer, I have put comments to help you understand how to solve the question.** This means that if you haven't got the right answer, you can read through what I've written and find out where you went wrong. This will help you get the right answer next time around in your exam.

Collins · do brilliantly!

Includes Tony Buzan's
Mind Maps®

2005 Exam **Practice**

GCSE Maths

Exam practice at its **best**

Paul Metcalf

Series Editor: Jayne de Courcy

William Collins' dream of knowledge for all began with the publication of his first book in 1819.
A self-educated mill worker, he not only enriched millions of lives, but also founded a flourishing publishing house.
Today, staying true to this spirit, Collins books are packed with inspiration, innovation and practical expertise.
They place you at the centre of a world of possibility and give you exactly what you need to explore it.

Collins. Do more.

Published by Collins
An imprint of HarperCollins*Publishers*
77–85 Fulham Palace Road
Hammersmith
London
W6 8JB

Browse the complete Collins catalogue at
www.collinseducation.com

First published 2001
This revised edition published 2005

10 9 8 7 6 5 4 3 2 1

ISBN 0 00 719494 3

British Library Cataloguing in Publication Data
A Catalogue record for this publication is available from the British Library

Acknowledgements
The Author and Publishers are grateful to the following for permission to reproduce copyright material:
AQA/NEAB The following questions: p.13 Q2, p.15 Q5, p.16 Q1, p.31 Q2, p.40 Q1, p.46 Q1, p.54 Q5, p.90 Q2, p.92 Q5 are reproduced with the permission of the Assessment and Qualifications Alliance.
AQA/SEG The following questions: p.8 Q1, p.18 Q3, p.26 Q2, p.26 Q5, p.30 Q3, p.37 Q1, p.39 Q4, p.42 Q2, p.45 Q2, p.45 Q5, p.53 Q2, p.53 Q3, p.57 Q3, p.59 Q2, p.61 Q2, p.62 Q4, p.67 Q2, p.74 Q2, p.76 Q2, p.79 Q3, p.83 Q1 are reproduced with the permission of the Assessment and Qualifications Alliance.
CCEA The following questions: p.18 Q1, p.39 Q1, p.58 Q4, p.59 Q1 are reproduced with the permission of the Northern Ireland Council for the Curriculum Examinations.
Edexcel The following questions: p.22 Q4, p.36 Q2, p.37 Q2, p.53 Q1, p.63 Q7, p.79 Q4 are reproduced with the permission of Edexcel.
OCR The following questions: p.8 Q2, p.10 Q1, p.10 Q3, p.15 Q4, p.27 Q2, p.33 Q1, p.43 Q2, p.48 Q4, p.54 Q4, p.54 Q7, p.55 Q1, p.61 Q1, p.61 Q3, p.63 Q6, p.67 Q1, p.68 Q4, p.73 Q1, p.76 Q3, p.79 Q2, p.79 Q6, p.85 Q4, p.92 Q2, p.92 Q3 are reproduced with the permission of Oxford Cambridge RSA Examinations.
WJEC The following questions: p.15 Q1, p.16 Q2, p.19 Q1, p.22 Q5, p.27 Q1, p.30 Q2, p.33 Q3, p.36 Q1, p.39 Q5, p.64 Q1, p.69 Q5, p.72 Q2, p.77 Q2, p.85 Q9, p.86 Q1, p.89 Q7, p.90 Q1 are reproduced with the permission of the Welsh Joint Education Committee/Cyd-bwyllgor Addysg Cymru.
Note: The Awarding Bodies listed above accept no responsibility whatsoever for the accuracy or method of working in the answers given, which are solely the responsibility of the author and publishers.

Every effort has been made to contact the holders of copyright material, but if any have been inadvertently overlooked, the Publishers will be pleased to make the necessary arrangements at the first opportunity.

Edited by Karen Westall and Joan Miller
Mind Maps artwork by Kathy Baxendale
Production by Katie Butler
Book design by Bob Vickers
Printed and bound by Printing Express, Hong Kong

All about your Maths exam

Written papers and calculators

The majority of GCSE Maths exams are assessed on the basis of **two written** papers. You will be expected to answer one of these papers using a calculator and the other paper without a calculator.

Some of the questions in this book should be answered without the use of a calculator. They have this symbol by them.

Some questions would be very difficult to answer without using a calculator, or specifically need to be answered with the use of a calculator. They have this symbol by them.

In general, though, when you are using this book, it is a good idea to **practise answering questions with and without the use of a calculator**.

Exam Tips

- **Always read the question carefully** and make sure that you answer all of the parts.
- **Always show your working** and write down your calculations.
- **Make sure that your answer is reasonable** and appropriate to the question set.
- **Don't cross out any work** until you have replaced it by something better.
- If you get stuck on a question, then miss it out and **come back to it later on**.
- **Make sure you haven't missed out** any pages by mistake (especially the back page).
- When you have finished **go back and check through your work again**.
- **Recheck your answer to question 1** – this is often answered wrongly the first time.

The topics covered by your specification

The content and assessment of the different Exam Boards' GCSE specifications are very similar and usually differ only in the way in which coursework is assessed.

This book is divided into 23 chapters covering the most common mathematical topics in the GCSE Mathematics specifications:

Shape, Space and Measures	Chapters 1–6
Number	Chapters 7–11
Handling Data	Chapters 12–17
Algebra	Chapters 18–23

I have selected the questions from across all Exam Boards. Have a go at all of them as plenty of practice really will help you to achieve a high grade.

Tiers of entry

All GCSE Maths exams are offered at **three tiers** and the available grades are as follows:

Tier of entry	Available grades
Foundation tier	D, E, F, G and U only
Intermediate tier	B, C, D, E and U only
Higher tier	A*, A, B, C and U only

The written papers assess the full range of grades for the tier of entry. The papers consist of **questions of varying lengths** and you will be expected to answer all of the questions on the paper itself.

Intermediate or Higher?

Some of the chapters in this book are aimed at students sitting Intermediate papers and some at those sitting Higher papers. If you are entered for the Intermediate tier, then you do not need to cover the work for the Higher tier which is identified by a large  at the top of each page in the chapter.

Why MIND MAPS® will help you

Introduction by TONY BUZAN

Mind Maps really are a shortcut to exam success. They can help you to plan your revision, to organise your work and to remember important information in your exams.

What is a Mind Map?

First of all, a Mind Map is a way of organising your thoughts. Then, because you have created it yourself, and used colours and pictures to make it 'stick' in your brain, it is a way of helping you to remember those thoughts.

A Mind Map is like exercise for your brain. The shapes, colours and pictures on your Mind Map keep your brain active and help you to remember the important facts and ideas.

How do I create a Mind Map?

An example Mind Map for **Transformations** is shown below. This shows you some of the important rules to follow:

- Start with paper, lots of coloured pens, and The Big Idea.
- Turn the paper sideways and write the name of a subject – or the big idea – in the middle of the paper. Add a drawing to help you remember. You don't have to be a great artist: the important thing here is to keep things lively and colourful.
- Now let your imagination get to work! Draw 4 or 5 branches from the central idea for each of the main topics. Write the key word for that topic on the branch, filling up the length of the branch.
- Then you can add smaller branches as you think of how each main topic can break down further. Use only one word for each idea.
- Use a different colour for each main branch and its sub-branches. Colour adds fun and will help you remember.
- Illustrate the branches with small pictures, shapes and colours.
- You can use arrows, underlining or boxes, to link ideas on different branches.

Not just facts and figures

A Mind Map is about much more than remembering facts, figures, vocabulary, names, dates and places. In Maths, you can use them to remember formulae, to record and remember data and to sort information. You can also use them in all your subjects to help plan your answers and to remember the best way to answer questions.

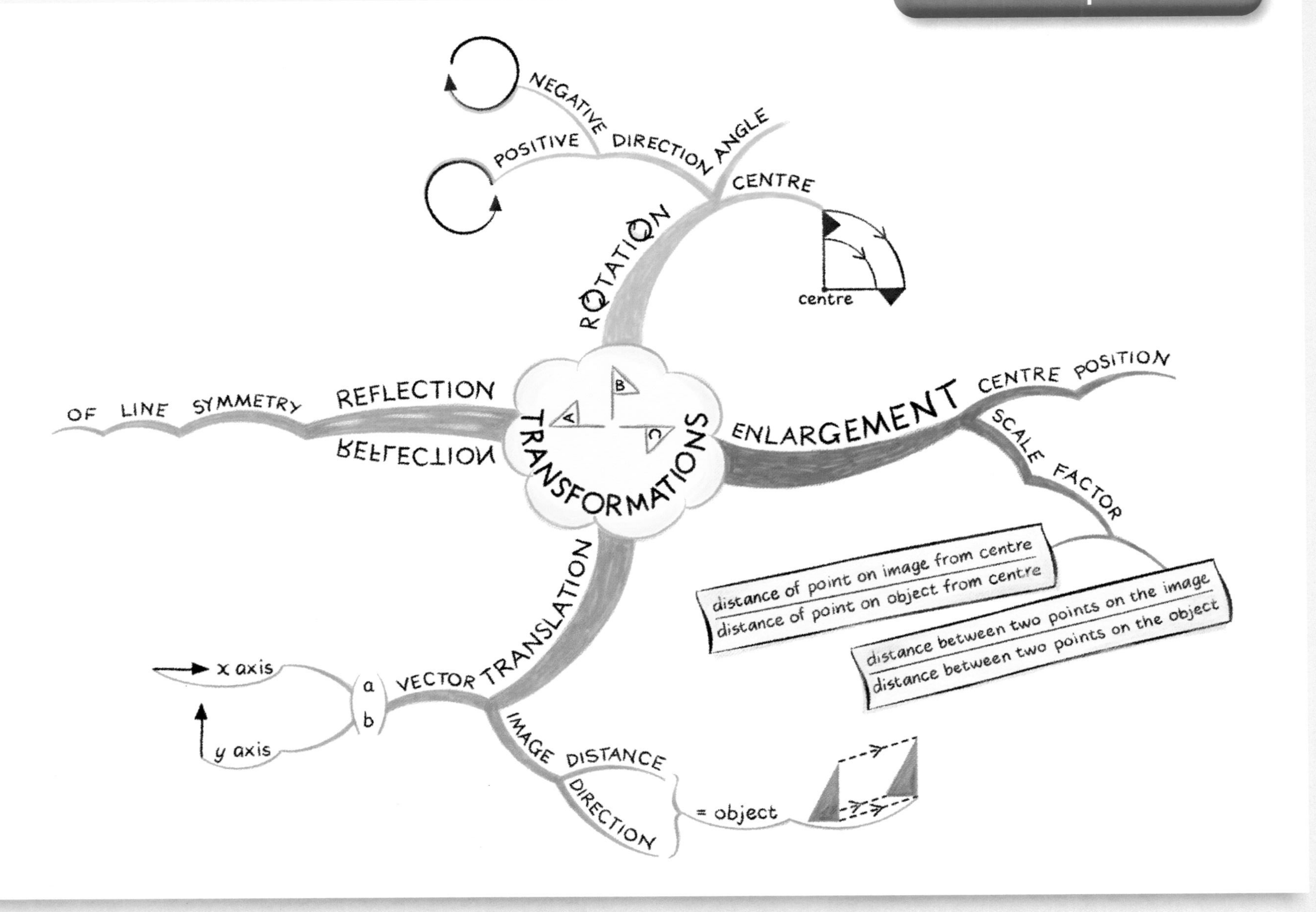
TRANSFORMATIONS
A
B
C
ROTATION
ANGLE
DIRECTION
POSITIVE
NEGATIVE
CENTRE
centre
ENLARGEMENT
CENTRE POSITION
SCALE FACTOR
distance of point on image from centre
distance of point on object from centre
distance between two points on the image
distance between two points on the object
TRANSLATION
VECTOR
a
b
x axis
y axis
IMAGE
DISTANCE
DIRECTION
= object
REFLECTION
REFLECTION
SYMMETRY
LINE
OF

1 Transformations

Exam Questions and Answers

1

1 The diagram shows three triangles P, Q and R.

(a) Describe fully the single transformation which takes P onto Q.

(b) Describe fully the single transformation which takes P onto R.

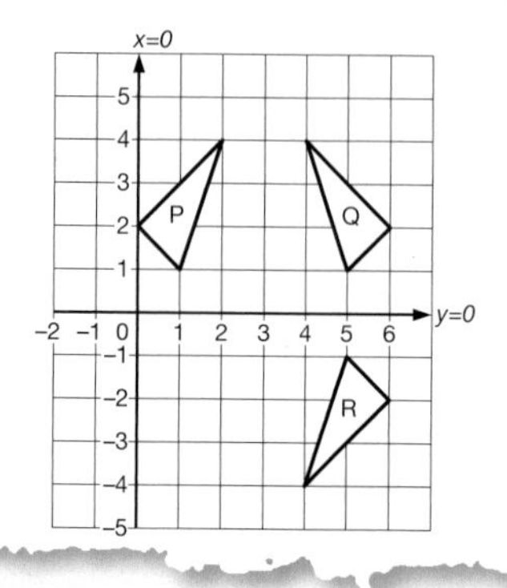

1

Answer

(a) *Reflection in $x = 3$*

(b) *Rotation of 180° about the point (3, 0)*

How to score full marks

- The single transformation is a reflection.
- You should have noticed that all the coordinates on this line have an x-value of 3, so the equation of the line is $x = 3$.
- Here, the single transformation is a rotation.
- The rotation is 180°. You don't need to say clockwise or anticlockwise as the result is the same in either case.
- The centre of rotation is (3, 0); you must state this for full marks.

2

2 (a) Describe fully the transformation that maps triangle T onto triangle A.

(b) Reflect triangle T in the line $y = x$ and label the image B.

(c) Rotate triangle T through 90° anticlockwise about (0, 1) and label the image C.

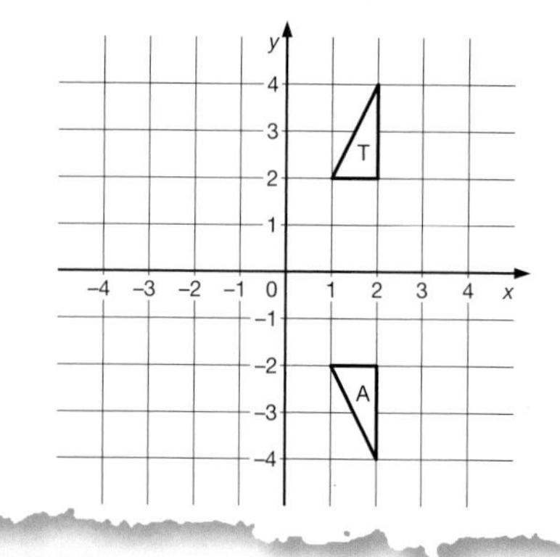

2

Answer

(a) *Reflection in the x-axis (or the line $y = 0$).*

(b) *See the diagram opposite.*

(c) *See the diagram opposite.*

How to score full marks

- The transformation is a reflection in the x-axis, or you can describe it as the line $y = 0$.
- Both transformations are shown on this diagram.

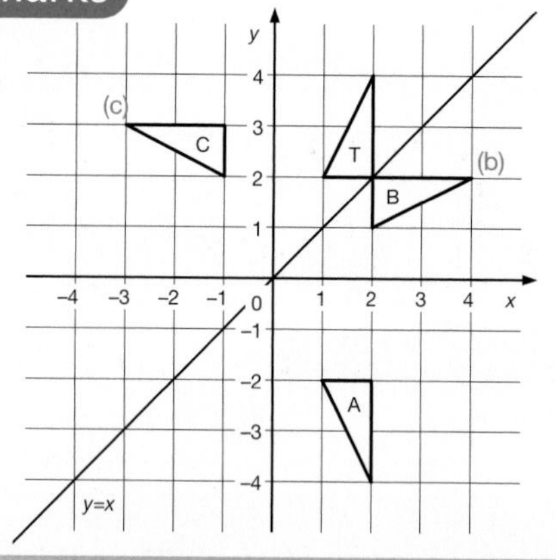

Key points to remember

Object and image

The object is the point or collection of points before the transformation. The image is the point or collection of points after the transformation.

Subsequent images of points ABC … are usually written as A'B'C'…, A"B"C" … or else $A_1B_1C_1$…, $A_2B_2C_2$ … .

Different transformations

You need to be able to recognise and describe these transformations:

- reflection
- rotation
- enlargement
- translation

Reflection

In a reflection, any two corresponding points on the object and image are the same distance away from a fixed line (called the line of symmetry or mirror line), but on opposite sides of it.

Define a reflection by giving the position of the line of symmetry.

Rotation

In a rotation, lines between any two corresponding points on the object and on the image meet at a fixed point (called the centre of rotation) and they always make the same angle.

Define a rotation by giving the angle (or turn) and direction of the rotation along with the position of the centre of rotation.

Enlargement

In an enlargement, the distance between a point on the image and a fixed point (called the **centre of enlargement**) is a factor or multiple of the distance between the corresponding point on the object and the fixed point.

Define an enlargement by giving the scale factor and the position of the centre of enlargement.

$$\text{scale factor} = \frac{\text{distance of point on image from centre}}{\text{distance of corresponding point on object from centre}}$$

or

$$\text{scale factor} = \frac{\text{distance between two points on the image}}{\text{distance between two corresponding points on the object}}$$

Translation

In a translation, the distance and the direction between corresponding points on the object and image are the same.

Define a translation by giving the distance and direction of the translation, usually in the **vector form** $\begin{pmatrix} a \\ b \end{pmatrix}$ where a is the distance in the x-direction and b is the distance in the y-direction.

DON'T MAKE THESE MISTAKES …

✗ **Combinations of transformations**

Don't forget that **two or more transformations** can usually be **combined** as a single transformation. If the question asks for the **single transformation** then you must identify **one transformation** and **describe it fully**.

Describe it fully means:

Transformation	*Provide information on:*
Reflection	line of reflection
Rotation	angle + direction + centre of rotation
Enlargement	scale factor + centre of enlargement
Translation	units across and units up (usually as a vector)

Questions to try

1

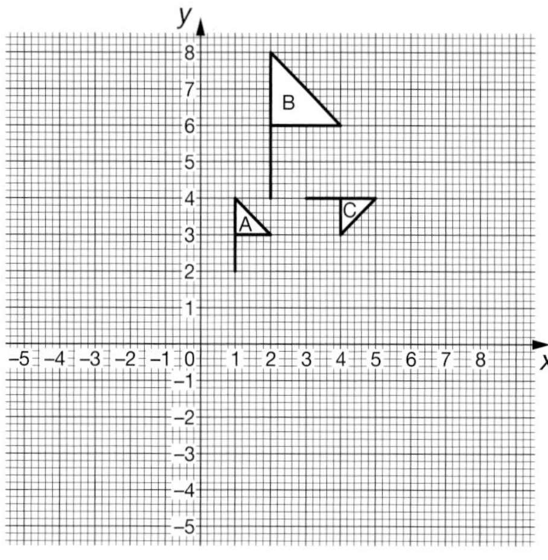

(a) Draw the image of the flag A after a translation of $\begin{pmatrix} -4 \\ 3 \end{pmatrix}$. Label it T.

(b) Describe fully the single transformation that will map:
   (i) A onto B   (ii) A onto C.

2

(a) Reflect the triangle ABC in the y-axis. Label the reflection A'B'C'.

(b) Reflect the triangle A'B'C' in the line $y = x$. Label the reflection A"B"C".

(c) Describe fully the single transformation which maps the triangle A"B"C" back onto ABC.

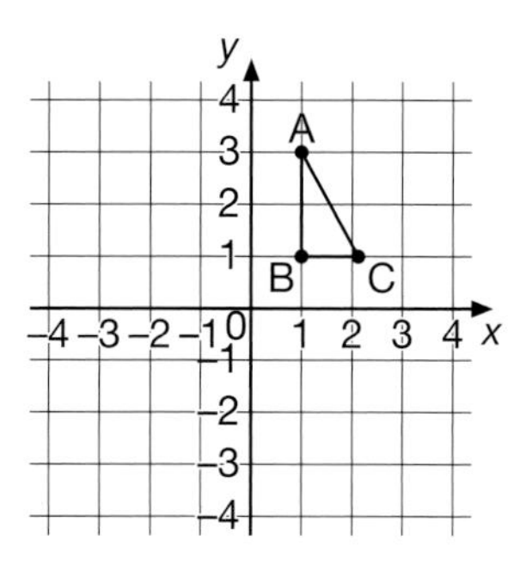

3

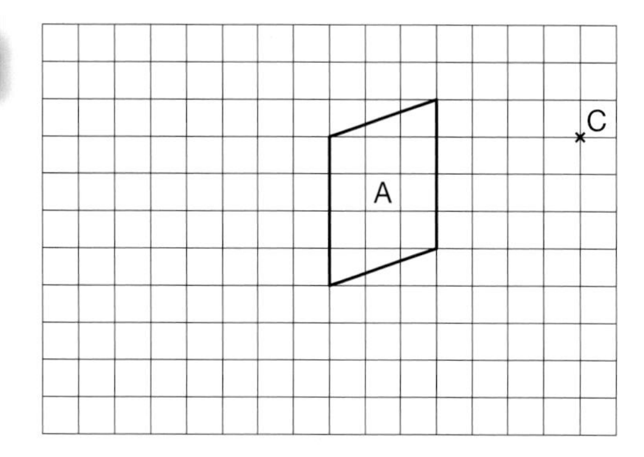

Parallelogram A is drawn on the grid above.

(a) Enlarge A with centre of enlargement C and scale factor 2.

(b) (i) Calculate the area of A.
   (ii) Calculate the area of the enlarged parallelogram.
   (iii) What is the ratio of these areas in the form $1 : n$?

4

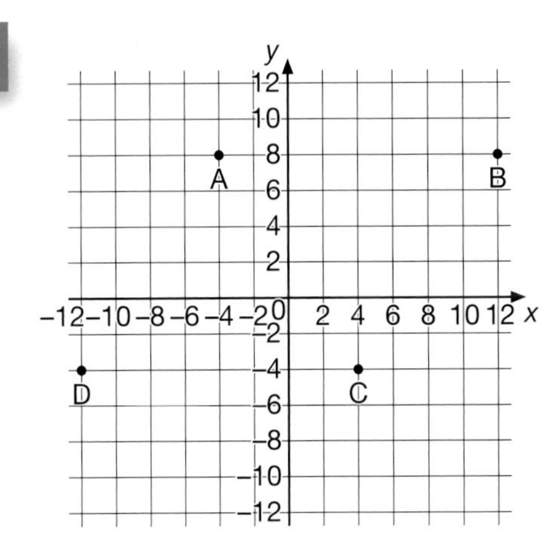

An enlargement scale factor $\frac{1}{4}$ and centre (0, 0) transforms parallelogram ABCD onto $A_1B_1C_1D_1$. The parallelogram $A_1B_1C_1D_1$ is translated by vector $\begin{pmatrix} 3 \\ -2 \end{pmatrix}$ onto $A_2B_2C_2D_2$.

What are the coordinates of the points A_2, B_2, C_2 and D_2?

5

(a) Rotate the rectangle P through an angle of 90° clockwise about centre of rotation (0, 0). Label the rotation Q.

(b) Rotate the rectangle Q through an angle of 90° anticlockwise about centre of rotation (3, 2). Label the rotation R.

(c) Describe fully the single transformation which maps the rectangle P onto R.

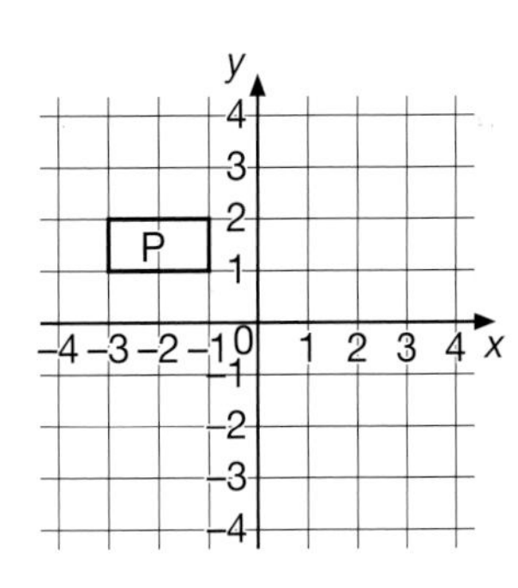

The answers can be found on pages **93–94**.

2 Pythagoras' theorem

Exam Questions and Answers

1

1 The dimensions of the rectangle are shown in the diagram.
Calculate the length of the diagonal. Leave your answer in surd form.

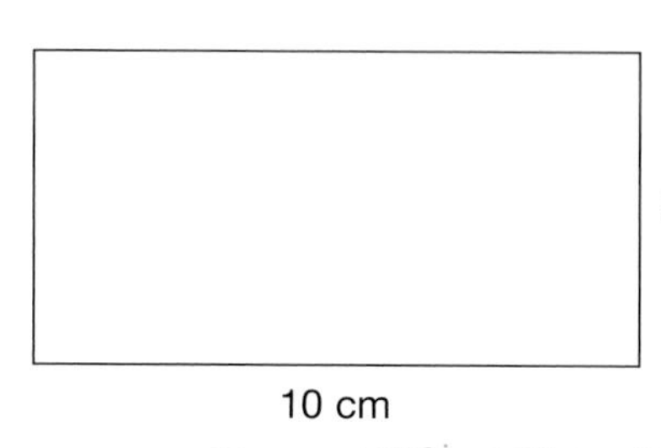

Diagram not to scale

1

Answer

$x^2 = 5^2 + 10^2$
$x^2 = 25 + 100$
$x^2 = 125$
$x = \sqrt{125}$
Diagonal = $\sqrt{125}$ cm

How to score full marks

- Start by drawing a diagram with the diagonal identified.

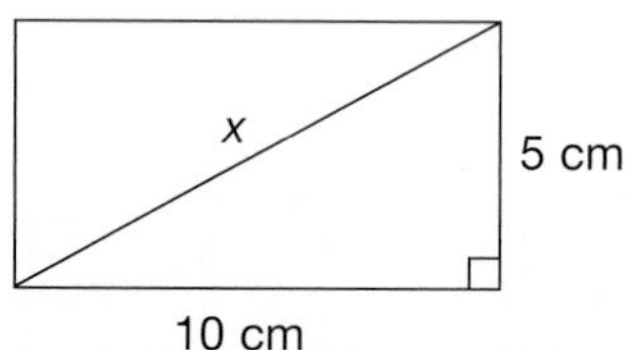

- Use Pythagoras' theorem to find the length of the hypotenuse (diagonal).
- Square the numbers and remember
  $5^2 = 5 \times 5 = 25$ (not 10)
  $10^2 = 10 \times 10 = 100$ (not 20).
- Take the square root on both sides to find the length.
- Leave your answer as a surd (i.e. as a square root). You could simplify $\sqrt{125}$ to get $\sqrt{125} = \sqrt{5 \times 5 \times 5} = 5\sqrt{5}$.

Note: On a calculator paper you may be asked to give your answer to an appropriate degree of accuracy (say 3 significant figures).

Your answer will look like this.

$x^2 = 125$
$x = \sqrt{125}$
$x = 11.180\,339\,887\,5$
$x = 11.2$ (3 s.f.)
Diagonal = 11.2 cm (3 s.f.)

2 An oil rig is 15 kilometres east and 12 kilometres north from Kirrin.

(a) Calculate the direct distance from Kirrin to the oil rig. (The distance is marked x on the diagram.)

(b) An engineer flew 14 kilometres from Faxtown to the oil rig. The oil rig is 10 kilometres west of Faxtown. Calculate how far south the oil rig is from Faxtown. (The distance is marked y on the diagram.)

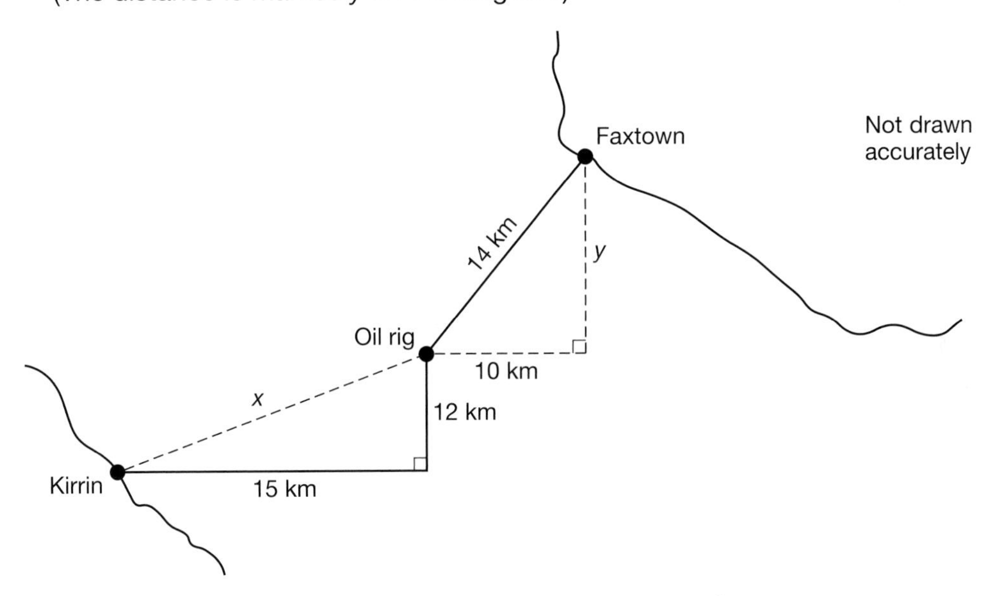

Answer

(a) $x^2 = 12^2 + 15^2$

$x^2 = 144 + 225$

$x^2 = 369$

$x = \sqrt{369}$

$x = 19.209373$

$x = 19.2$ km (1 d.p.)

(b) $14^2 = y^2 + 10^2$

$196 = y^2 + 100$

$196 - 100 = y^2$

$96 = y^2$

$y^2 = 96$

$y = \sqrt{96}$

$y = 9.7979590$

$y = 9.8$ km (1 d.p.)

How to score full marks

- You should start by using Pythagoras' theorem to find the length of the hypotenuse (marked x).
- Square the numbers you know, add them and then take the square root of both sides.
- Round the answer to a suitable degree of accuracy and remember to include the units of distance.
- The second part of the question uses Pythagoras' theorem, but take care as the length (marked y) is **not** the hypotenuse. This sort of question is popular on exam papers!
- Subtract 100 from both sides to isolate the y^2 term, then simplify.
- Place the y^2 on the left-hand side (to make y^2 the subject).
- Take the square root of both sides.
- Round the answer to a suitable degree of accuracy and remember to include the units of distance.

Key points to remember

Right-angled triangles

For any right-angled triangle, the square of the length of the hypotenuse is equal to the sum of the squares of the lengths on the other two sides.

$c^2 = a^2 + b^2$

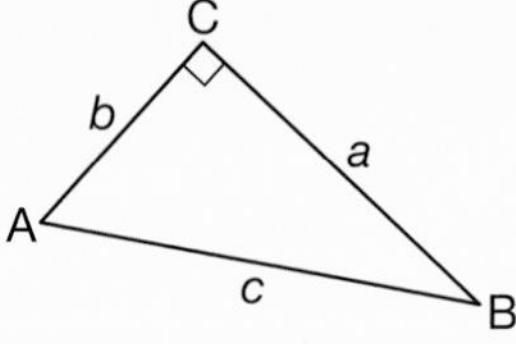

For a right-angled triangle, the side opposite the right angle is called the hypotenuse and this is always the longest side.

If you are given the hypotenuse, use $c^2 = a^2 + b^2$

If you are not given the hypotenuse, use $c^2 = a^2 + b^2$ and rearrange as:

$a^2 = c^2 - b^2$ or $b^2 = c^2 - a^2$

Pythagorean triples

The set of numbers:
3, 4, 5 is called a Pythagorean triple because $3^2 + 4^2 = 5^2$
5, 12, 13 is called a Pythagorean triple because $5^2 + 12^2 = 13^2$

Pythagorean triples are **sets of three whole numbers** that satisfy the rule for Pythagoras' theorem.

If you know your Pythagorean triples then you can solve right-angled triangles easily …

Remember 3, 4, 5 and 5, 12, 13… and also any multiples of them:

3, 4, 5
6, 8, 10 are all this shape:
9, 12, 15 …

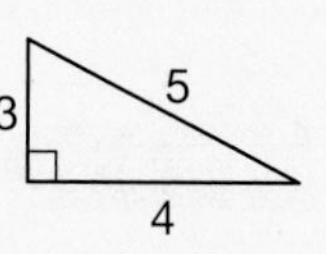

5, 12, 13
10, 24, 26 are all this shape:
15, 36, 39 …

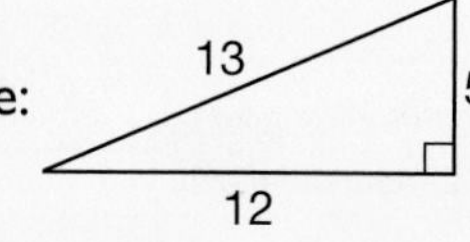

DON'T MAKE THESE MISTAKES …

✗ Remember that **calculate** means you **must calculate** the answer, so **don't just measure** it from the diagram.

This is particularly true where the diagram has any of the following notes written next to it:

Diagram not drawn to scale
Diagram NOT accurately drawn
Not to scale
Not drawn to scale
Not drawn accurately

✗ Remember that x^2 **means** $x \times x$ (it does **not** mean $x \times 2$).

So, for example, $3^2 = 3 \times 3$, $4^2 = 4 \times 4$, $5^2 = 5 \times 5$, $6^2 = 6 \times 6$ ...

although 2^2 is equal to 2×2 which (presumably) causes the problems!

✗ Remember that on a non-calculator paper you may need to leave your answer in surd form – but check you can't simplify – $\sqrt{25} = 5$, $\sqrt{121} = 11$.

✗ **Careful!**
These numbers do NOT form a Pythagorean triple.

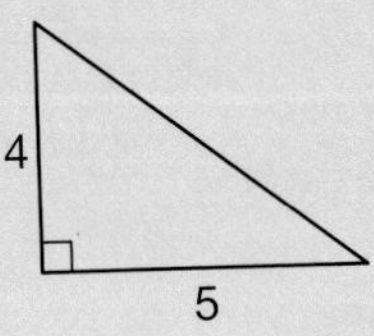

Questions to try

1 The diagram on the right shows the layout of five paths in the garden.

AB is 8.3 m long. BC is 6.1 m long. BD is 5.0 m long. ABC and BDA are both right angles.

(a) Calculate the length of the path AC, giving your answer to a suitable degree of accuracy.

(b) Calculate the length of the path AD, giving your answer to a suitable degree of accuracy.

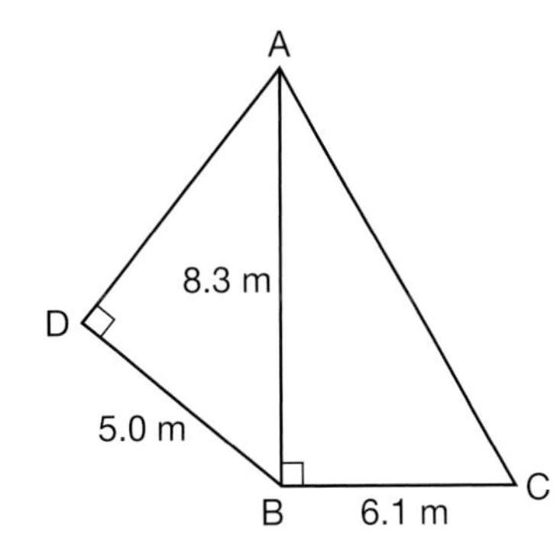

2 Calculate the length of a diagonal of a square of side 10 cm. Leave your answer in surd form.

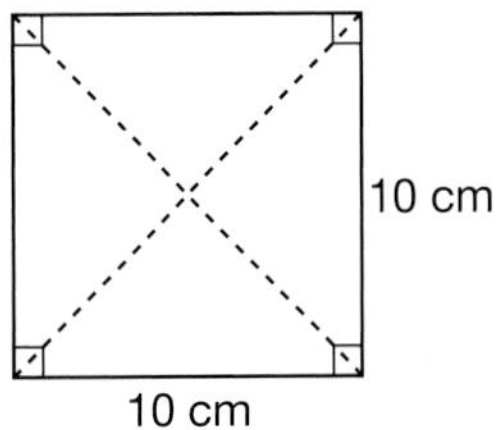

3 Calculate the height of this isosceles triangle. Leave your answer in surd form.

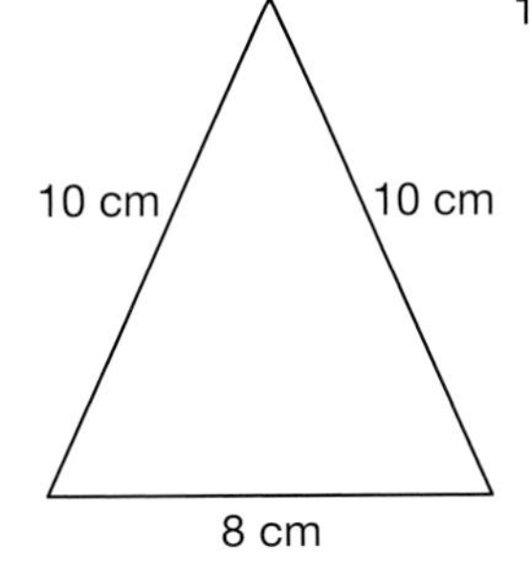

4 This is a parallelogram. The lengths marked are in centimetres.

(a) Work out the area of the parallelogram. State the units of your answer.

(b) Work out the perimeter of the parallelogram. State the units of your answer.

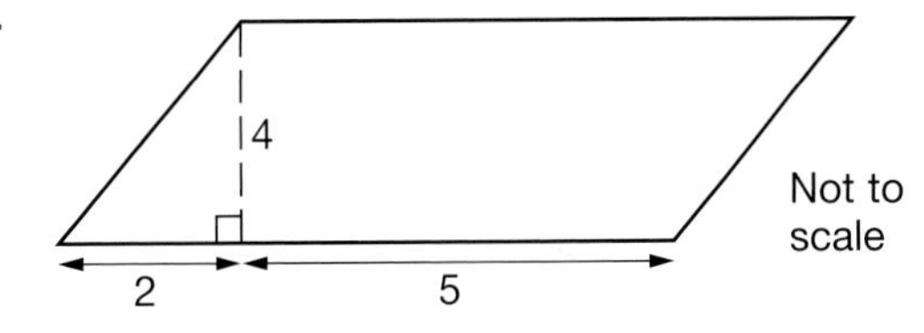

5 A ladder, 2.75 m long, leans against a wall.

The bottom of the ladder is 1.80 m from the wall, on level ground.

Calculate how far the ladder reaches up the wall.

Give your answer to an appropriate degree of accuracy.

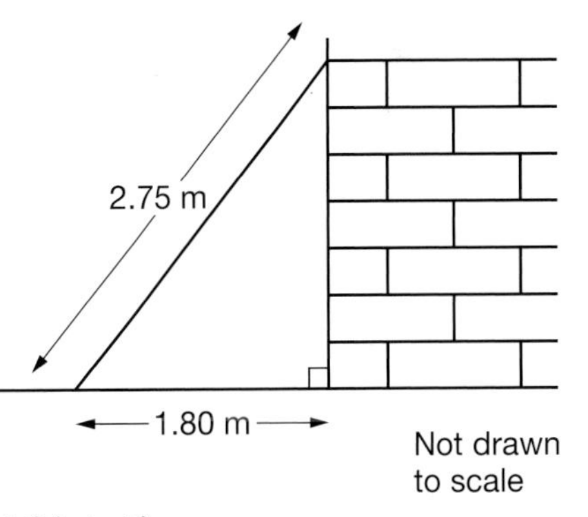

6 The triangle ABC has vertices at A(3, –1), B(6, 1) and C(–1, 5). Find the lengths AB, BC and CA and prove that ABC is a right-angled triangle.

The answers can be found on pages **94–95.**

3 Angle and tangent properties of circles

Exam Questions and Answers

1

1 In the diagram, O is the centre of the circle.

Angle COA = 100°.

Calculate: (a) angle CBA (b) angle CDA.

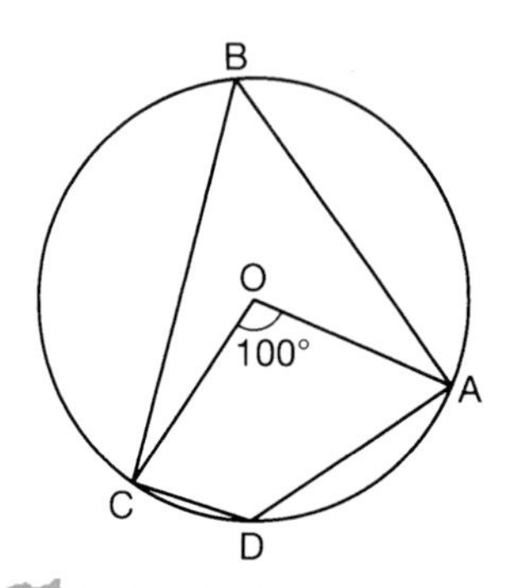

Not drawn accurately

1 **Answer**

(a) angle CBA = 50°

(b) angle CDA = 130°

► ∠CBA = 50° (as the angle subtended by an arc AC at the centre is twice that subtended at the circumference)

► ∠CDA = 130° (as the opposite angles of a cyclic quadrilateral ABCD are supplementary so they add up to 180°)

2

2 A, B, C and D are four points on the circumference of a circle centre O.

AT is a straight line passing through the centre of the circle.

The tangent PT meets the circle at D.

Given that $O\hat{A}D = 25°$, find **each** of the following angles.
Give reasons for your answer.

(a) $A\hat{B}C$ (b) $C\hat{O}D$
(c) $A\hat{D}T$

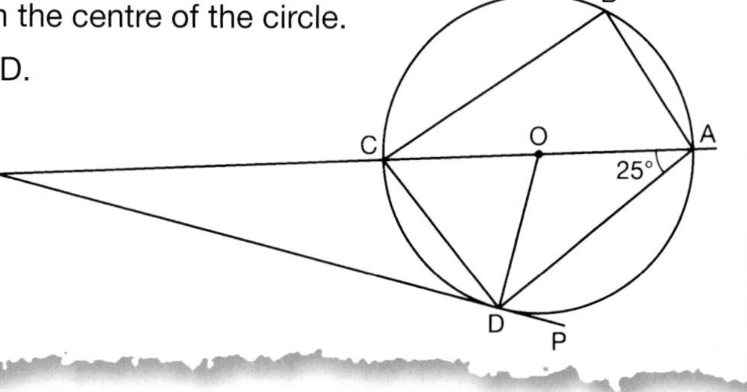

2 **Answer**

(a) $A\hat{B}C = 90°$

(b) $C\hat{O}D = 50°$

(c) $A\hat{D}T = 115°$

► $A\hat{B}C = 90°$ (as the angle in a semicircle is always 90°)

► $C\hat{O}D = 50°$ (as the angle subtended by an arc/chord CD at the centre, $C\hat{O}D$, is twice that subtended at the circumference, $C\hat{A}D$)

► $A\hat{D}O = 25°$ (base angles of isosceles triangle ODA)

► $O\hat{D}T = 90°$ (tangent is perpendicular to the radius)

► $A\hat{D}T = A\hat{D}O + O\hat{D}T = 25° + 90° = 115°$

Key points to remember

Questions on this subject are more likely to appear on the non-calculator paper.

Remember that it is always best to find angles in the order suggested by the question.

Don't forget to label the angles on the diagram as you find them.

Properties of angles in circles

- The angle subtended by an arc (or chord) at the centre is twice that subtended at the circumference.

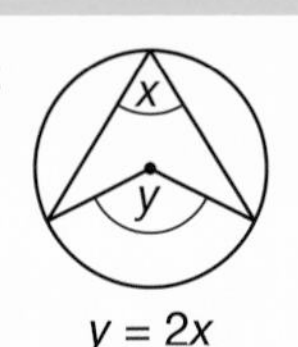

$y = 2x$

- The angle in a semicircle is always 90°.

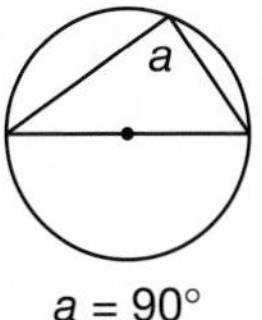

$a = 90°$

- Angles subtended by the same arc (or chord) are equal.

$p = q = r$

- The opposite angles of a cyclic quadrilateral are supplementary (they add up to 180°).

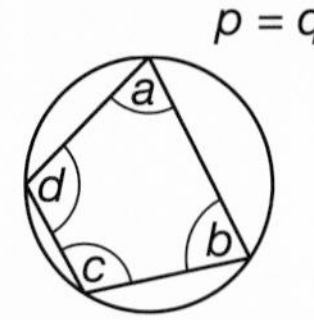

$a + c = 180°$
$b + d = 180°$

Properties of tangents to circles

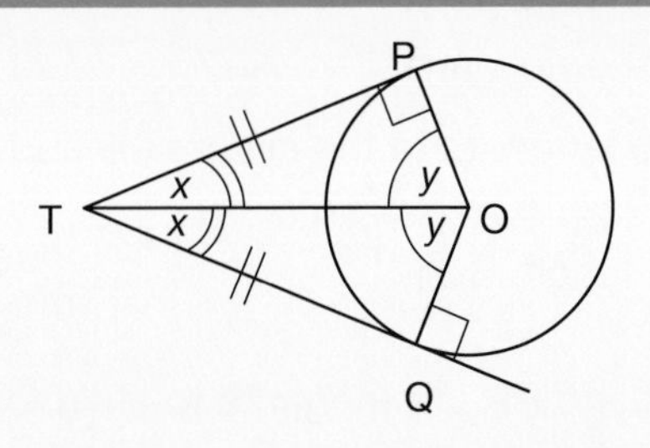

- A tangent to a circle is perpendicular to the radius at the point of contact.
- Triangles TPO and TQO are congruent:
  $\angle TPO = \angle TQO$, TO in both triangles
  PO = QO (radii of circle)
  so $\angle POT = \angle QOT$
  TP = TQ
- Tangents to a circle from an external point are equal in length.

Properties of chords in circles

- A perpendicular from the centre of a circle to a chord bisects the chord.

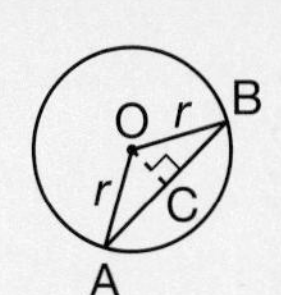

r is the radius of the circle

AC = CB

Conversely, a perpendicular bisector of a chord passes through the centre of the circle.

- Chords that are equal in length are equidistant from the centre of the circle.

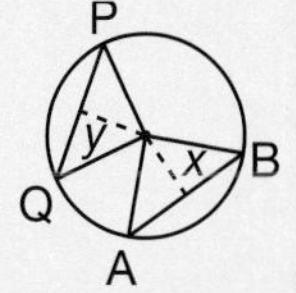

If AB = PQ then $x = y$

Conversely, chords that are equidistant from the centre of a circle are equal in length.

DON'T MAKE THESE MISTAKES ...

✗ For these questions you should always **use the properties given to find the angles – don't measure**, especially where the diagram has a note to say that it is not drawn to scale.

✗ **Never presume the values of angles**, always look for evidence ... the angle may look like a right angle but how can you be sure?

✗ Don't forget that the **radius of a circle** is always the same length, so watch out for **isosceles triangles**.

✗ You will be required to give a 'proof' of your answers ... so learn the properties given above.

Questions to try

1

O is the centre and AB is a diameter of the circle.

OD is parallel to BC and angle BAC = 56°.

Calculate the size of:

(a) angle ABC

(b) angle AOD

(c) angle ACD

(d) angle OAD.

The tangents to the circle at B and C meet at T.
Calculate the size of:

(e) angle BOT.

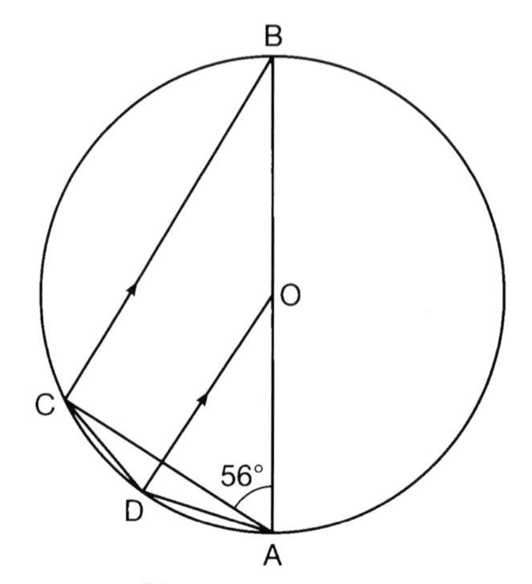

Diagram not drawn accurately

2

PQ and PR are tangents to a circle. Their points of contact are Q and R respectively.

O is the centre of the circle and angle POR = 75°.

Find:

(a) ∠POQ

(b) ∠OPQ

(c) ∠OPR.

Give reasons for your answers.

R
75°
O
P
Q

Not to scale

3

AB, BC and CA are tangents to the circle at P, Q and R respectively.

The centre of the circle is at O.

AC is parallel to PO and angle ACB is 40°.

Work out angle POQ.

Give reasons for your answer.

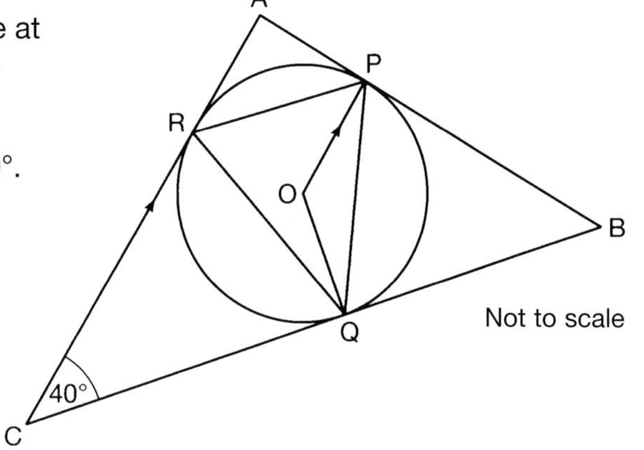

Not to scale

4

A chord AB is drawn on a circle of radius 6 cm. If the perpendicular distance of the chord from the centre of the circle is 4 cm, calculate the length of the chord.

5

Two chords PQ and RS are parallel to each other, on opposite sides of a circle of radius 15 cm. If PQ = 10 cm and RS = 12 cm, find the distance between the chords.

The answers can be found on pages **96–97.**

Exam Questions and Answers

1 The diagram shows a vertical pole, AB, standing on horizontal ground DBC.

The pole is held by two wires AC and AD.

The wire AC is 16 m long and makes an angle of 54° with the ground.

(a) Calculate the length of the pole AB. Give your answer to an appropriate degree of accuracy.

(b) The distance CD is 25 m. Calculate the angle the wire AD makes with the ground.

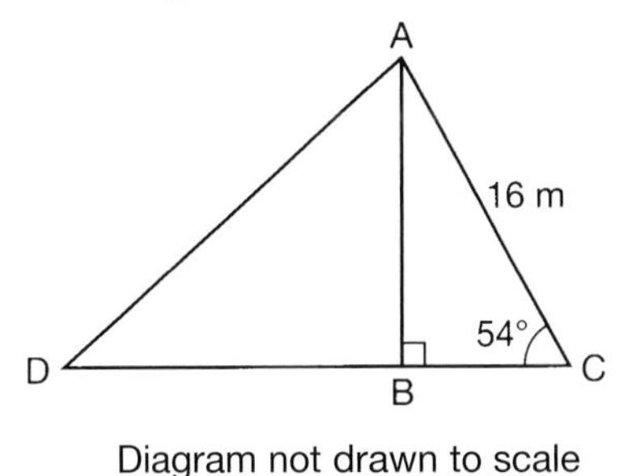

Diagram not drawn to scale

Answer

(a) $\sin 54° = \frac{opp}{hyp}$

$\sin 54° = \frac{AB}{16}$

$AB = 16 \times \sin 54°$

$AB = 12.944\,271\,91$

$AB = 12.9$ m (3 s.f.)

(b) $\cos 54° = \frac{adj}{hyp}$

$\cos 54° = \frac{BC}{16}$

$BC = 16 \times \cos 54°$

$BC = 9.404\,564\,037$

$BD = DC - BC$

$BD = 25 - 9.404\,564\,037$

$BD = 15.595\,435\,96$

$\tan\theta = \frac{AB}{BD}$

$\tan\theta = \frac{12.944\,271\,91}{15.595\,435\,96}$

$\tan\theta = 0.830\,003\,851$

$\theta = 39.692\,803\,8...°$

$\theta = 39.7°$ (3 s.f.)

How to score full marks

- The identified sides are 'opposite' and 'hypotenuse' so use:

  $\text{sine } x = \frac{\text{opposite}}{\text{hypotenuse}}$

- Rearrange the equation to make AB the subject.
- Round to an appropriate degree of accuracy (3 s.f. or 1 d.p. seems appropriate here).

Diagram not drawn to scale

- To calculate the angle ADB you need to know more about the triangle ADB. You have found that AB = 12.944 271 91 m
- You know that DC = 25 m but to find BD you need to work out BC first...
- The identified sides are 'adjacent' and 'hypotenuse' so use:

  $\text{cosine } x = \frac{\text{adjacent}}{\text{hypotenuse}}$

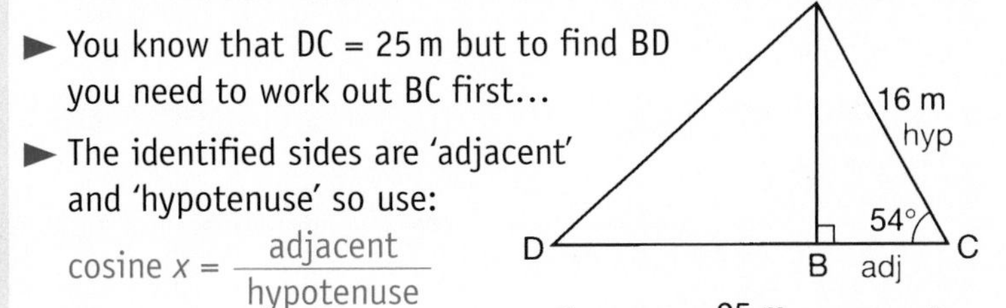

- Now to find ∠ ADB, the identified sides are 'adjacent' and 'opposite' so use:

  $\text{tangent } x = \frac{\text{opposite}}{\text{adjacent}}$

- Having found the value of tan θ, you can now use a calculator to find $\tan^{-1} 0.830\,003\,851$ (or arctan 0.830 003 851) to find the value of θ.
- Round the answer to an appropriate degree of accuracy.

2

2 The diagram shows the position of two beacons, A and B.

B is 4 km south and 6 km east of A.

Calculate the bearing of A from B.

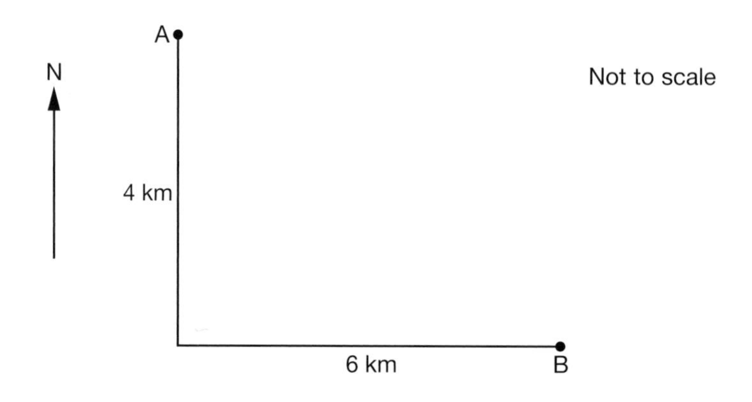

2

Answer

$\tan x = \frac{4}{6}$

$x = \tan^{-1}\frac{4}{6}$

$x = 33.690\,068°$

The bearing $= 270° + 33.690\,068°$

$= 303.690\,068°$

$= 304°$ (to the nearest degree)

How to score full marks

- To find the bearing you first need to find the angle marked x on the diagram.

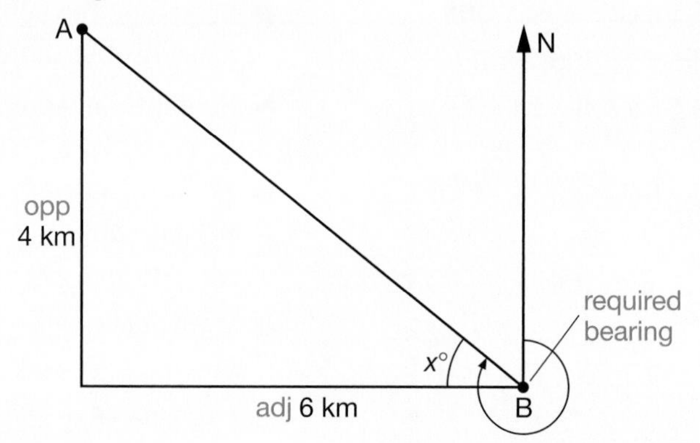

- The identified sides are 'adjacent' and 'opposite' so use:

  $\text{tangent } x = \frac{\text{opposite}}{\text{adjacent}}$

- Now you can use your calculator to find $\tan^{-1}\frac{4}{6}$ (or $\arctan\frac{4}{6}$) to find the angle.
- Finally, round to an appropriate degree of accuracy.

Key points to remember

The sides of a right-angled triangle are given special names, as in the diagrams.

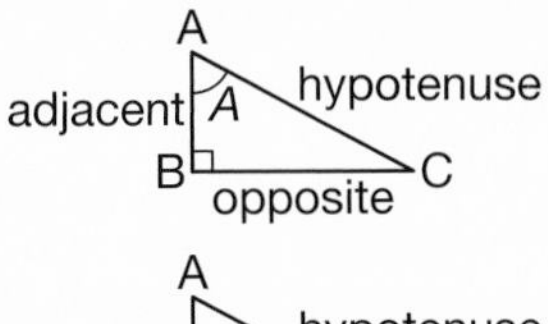

Of course the adjacent side and the opposite side depend on which angle you are considering, so it is important to label the sides you know and the sides you want to know.

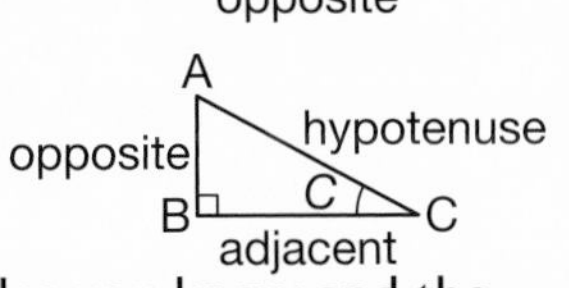

Sine of an angle

The sine of an angle (abbreviated as sin)

$$= \frac{\text{length of opposite side}}{\text{length of hypotenuse}}$$

$$\text{So } \sin A = \frac{\text{length of opposite side}}{\text{length of hypotenuse}} = \frac{BC}{AC}$$

$$\text{and } \sin C = \frac{\text{length of opposite side}}{\text{length of hypotenuse}} = \frac{AB}{AC}$$

Cosine of an angle

The cosine of an angle (abbreviated as cos)

$$= \frac{\text{length of adjacent side}}{\text{length of hypotenuse}}$$

$$\text{So } \cos A = \frac{\text{length of adjacent side}}{\text{length of hypotenuse}} = \frac{AB}{AC}$$

$$\text{and } \cos C = \frac{\text{length of adjacent side}}{\text{length of hypotenuse}} = \frac{BC}{AC}$$

Tangent of an angle

The tangent of an angle (abbreviated as tan)

$$= \frac{\text{length of opposite side}}{\text{length of adjacent side}}$$

$$\text{So } \tan A = \frac{\text{length of opposite side}}{\text{length of adjacent side}} = \frac{BC}{AB}$$

$$\text{and } \tan C = \frac{\text{length of opposite side}}{\text{length of adjacent side}} = \frac{AB}{BC}$$

Finding angles

You can reverse the process to find the angle, given two sides of a right-angled triangle.

If you are given the lengths of the sides, use the following ratios to find the angles.

$\sin^{-1} \frac{\text{opposite}}{\text{hypotenuse}}$ or $\arcsin \frac{\text{opposite}}{\text{hypotenuse}}$

$\cos^{-1} \frac{\text{adjacent}}{\text{hypotenuse}}$ or $\arccos \frac{\text{adjacent}}{\text{hypotenuse}}$

$\tan^{-1} \frac{\text{opposite}}{\text{adjacent}}$ or $\arctan \frac{\text{opposite}}{\text{adjacent}}$

Angles are frequently represented by Greek letters of the alphabet such as $\alpha, \beta, \gamma, \delta, \theta, \phi$ ....

Angles of elevation and depression

The angle of elevation is the angle up from the horizontal.

The angle of depression is the angle down from the horizontal.

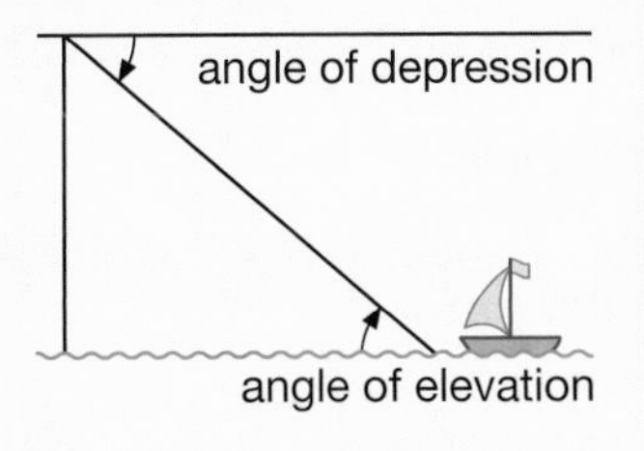

DON'T MAKE THESE MISTAKES ...

✗ **Most important! Don't forget to check that:**
- your calculator is set to DEG (degree) mode
- you are dealing with a right-angled triangle.

✗ Always check that your answers are **reasonable**:
- the **longest side** is always opposite the **right angle**
- the **smallest side** is opposite the **smallest angle**.

✗ Don't muddle 'adjacent' and 'opposite'. Identify the angle you are using and then **label the opposite and adjacent sides** (these will depend upon the identified angle).

✗ **Remember that the angles of a triangle add up to 180°, so don't waste time using sine, cosine and tangent to find the third angle.**

✗ For an **acute angle** (an angle between 0° and 90°), the **sine and the cosine of the angle are always less than 1**.

If your answer is not between 0 and 1, it is wrong!

Questions to try

1

In the diagram ABCD is a parallelogram.

BC = 5 cm and CD = 10 cm. Angle BCD = 48°.

Given that: sin 48° = 0.7431
cos 48° = 0.6691
tan 48° = 1.1106

calculate the area of the parallelogram, giving your answer to an appropriate degree of accuracy.

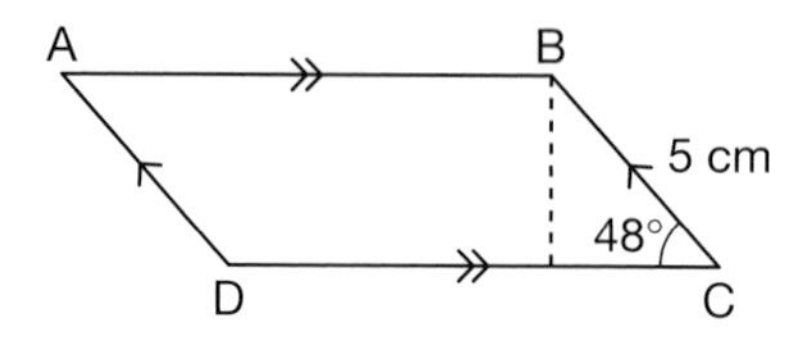

2

A rectangle measures 11 cm by 6 cm.

What angle does the diagonal make with the longest side?

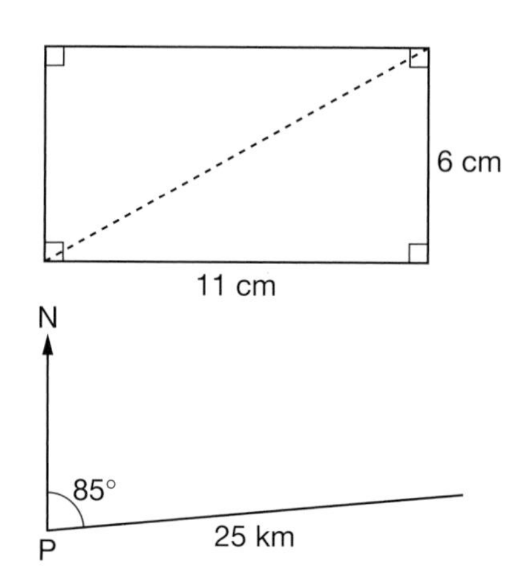

3

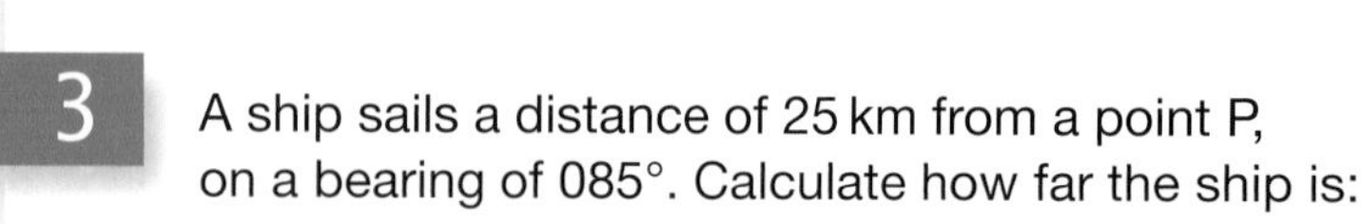

A ship sails a distance of 25 km from a point P, on a bearing of 085°. Calculate how far the ship is:

(a) to the north of P

(b) to the east of P.

4

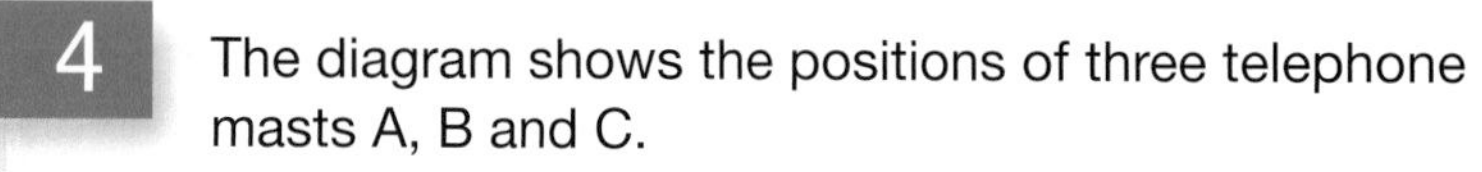

The diagram shows the positions of three telephone masts A, B and C.

Mast C is 5 kilometres due east of Mast B.

Mast A is due north of Mast B and 8 kilometres from Mast C.

(a) Calculate the distance of A from B.
Give your answer in kilometres, to three significant figures.

(b) (i) Calculate the size of the angle marked x°.
Give your angle correct to one decimal place.
(ii) Calculate the bearing of A from C.
Give your bearing correct to one decimal place.
(iii) Calculate the bearing of C from A.
Give your bearing correct to one decimal place.

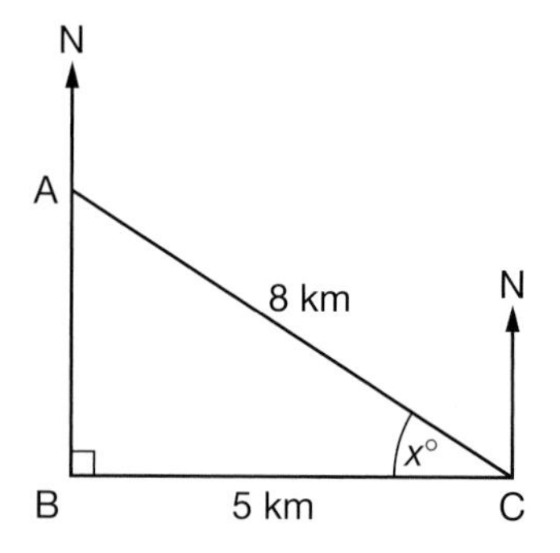

5

A boat B is moored 40 m from the foot of a vertical cliff. The angle of depression of the boat from the top of the cliff is 42°.

(a) Calculate the height of the cliff.

(b) The boat is released from its mooring and it drifts 250 m directly away from the cliff.

Calculate the angle of elevation of the top of the cliff from the boat.

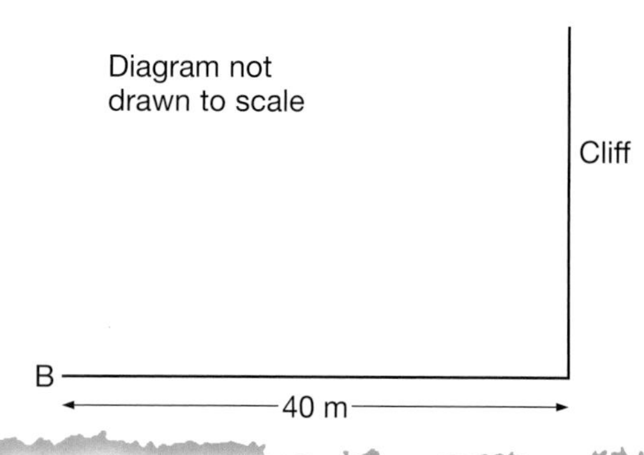

Exam Questions and Answers H

1 Find the size of angle A and the area of the triangle.

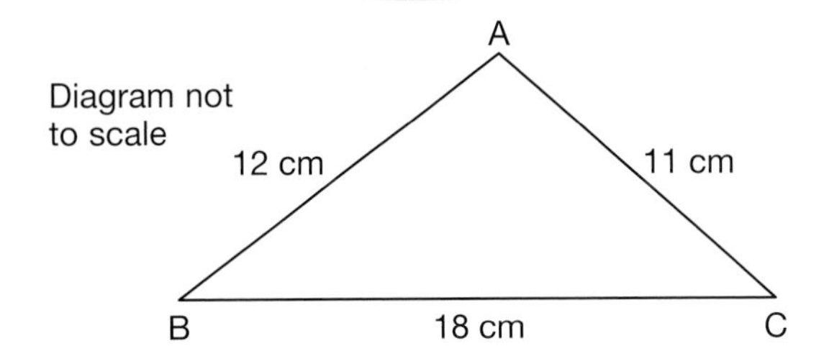

Answer

$\cos A = \dfrac{b^2 + c^2 - a^2}{2bc}$

$\cos A = \dfrac{11^2 + 12^2 - 18^2}{2 \times 11 \times 12}$

$\cos A = \dfrac{121 + 144 - 324}{264}$

$\cos A = \dfrac{-59}{264}$

$\cos A = -0.223\,484\,848$

$A = 102.913\,797\,8°$

$A = 102.9°$ (1 d.p.)

$\text{Area} = \frac{1}{2}bc\sin A$

$= \frac{1}{2} \times 11 \times 12 \times \sin 102.913\,797\,8°$

$= 64.330\,688\,63$

$= 64.3\text{ cm}^2$ (1 d.p.)

How to score full marks

- You are given three sides and need to find an angle, so you need to use the cosine rule.
  $\cos A = \dfrac{b^2 + c^2 - a^2}{2bc}$
- Substitute the given lengths, using the diagram to identify them.
- The negative value tells you that angle A is obtuse.
- Use the inverse button ($\cos^{-1}$ or arccos) on your calculator to find the angle.
- Round the answer to an appropriate degree of accuracy (usually 3 s.f. or 1 d.p.).
- To find the area of the triangle use:
  area of a triangle $= \frac{1}{2}bc\sin A$
- Remember to use the most accurate value of A (**not** the rounded value, as this is less accurate).
- Round to an appropriate degree of accuracy (3 s.f. or 1 d.p.).

2 Find the size of angle A in this triangle.

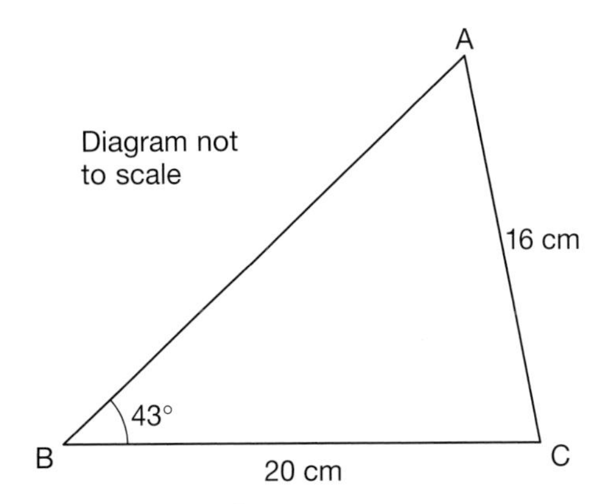

H

2

Answer

$$\frac{\sin A}{20} = \frac{\sin 43°}{16}$$

$$\sin A = \frac{20 \times \sin 43°}{16}$$

$\sin A = 0.852\,497\,95$

$A = 58.484\,408\,63°$ or

$A = 121.515\,591\,4°$

$A = 58.5°$ or $121.5°$ (1 d.p.)

How to score full marks

- In this case, since the information involves two sides and two angles then it is appropriate to use the sine rule so:

$$\frac{a}{\sin A} = \frac{b}{\sin B}$$

or $\frac{\sin A}{a} = \frac{\sin B}{b}$ by reciprocating both sides

- Multiply both sides by 20 to find an expression for sin *A*.
- Use the inverse button ($\sin^{-1}$ or arcsin) on your calculator.
- Don't forget that the value of $A = 58.484\,408\,63°$ is not unique. The graph shows that there is another possible value which satisfies $\sin A = 0.852\,497\,95$.

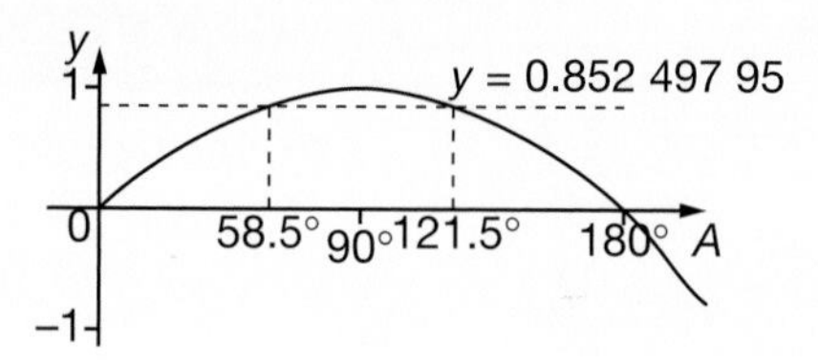

- The other value is $A = 121.515\,591\,4°$ (from the calculator).

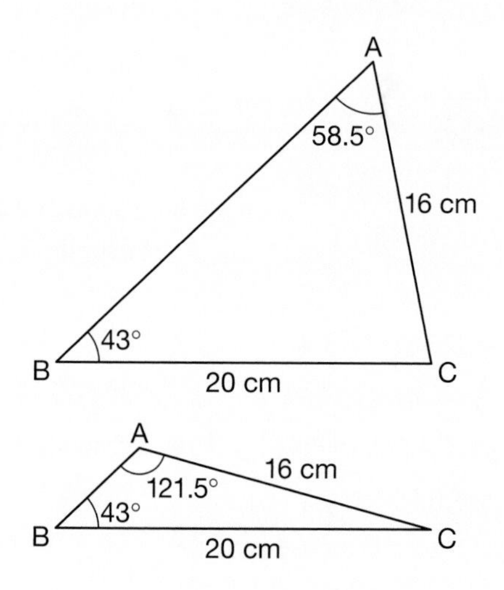

- You don't have this problem when you use the cosine rule as a negative value suggests an obtuse angle. When you use the sine rule, you can avoid it by finding the smallest angles first (if possible).

Key points to remember

Cosine rule

$a^2 = b^2 + c^2 - 2bc\cos A$

or $\cos A = \dfrac{b^2 + c^2 - a^2}{2bc}$

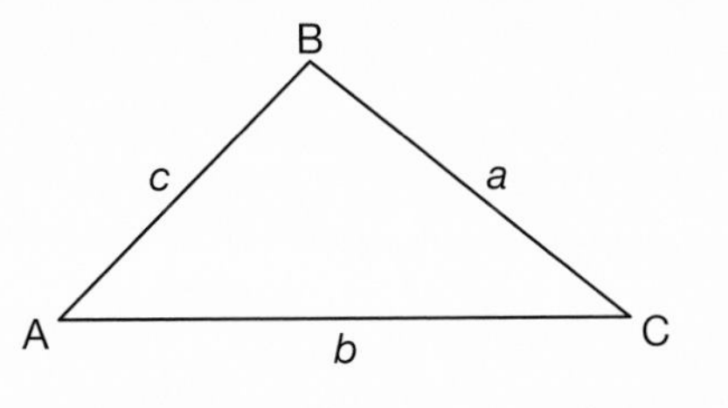

Sine rule

$\dfrac{a}{\sin A} = \dfrac{b}{\sin B} = \dfrac{c}{\sin C}$

or $\dfrac{\sin A}{a} = \dfrac{\sin B}{b} = \dfrac{\sin C}{c}$

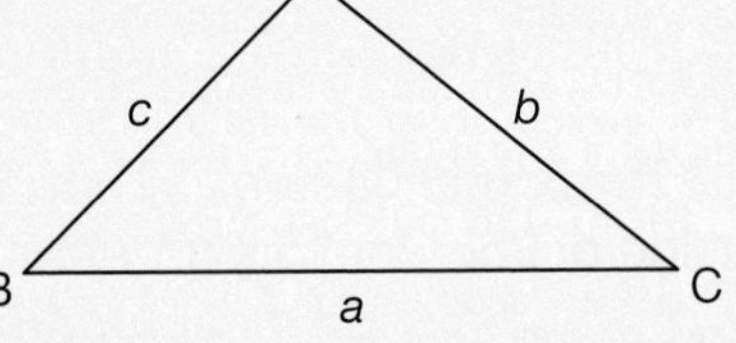

Area of a triangle

The sine and cosine rules can be extended to find the area of a triangle so that:

area of a triangle = $\frac{1}{2}ab\sin C$

where the angle C is the 'included' angle between sides a and b.

Similarly:

area of triangle = $\frac{1}{2}bc\sin A$

or area of triangle = $\frac{1}{2}ac\sin B$

A
c
b
B
a
C

When to use the rules

- Use the cosine rule when you are working with three sides and one angle.
- Use the sine rule when you are working with two sides and two angles.

DON'T MAKE THESE MISTAKES ...

✗ When you are using the **cosine rule**, be careful when you are subtracting $2bc\cos A$. **Work it out separately or use brackets**:
$a^2 = b^2 + c^2 - (2bc\cos A)$

✗ If the **cosine** of the angle is **negative**, the angle is **obtuse** (check that it is opposite the longest side).

✗ If the **sine** of the angle is **negative** you have probably made a **mistake**!

✗ When you are using the sine rule, don't forget that the largest angle might be **obtuse**. If one of the answers gives θ, the other answer is $180° - \theta$.

✗ **Don't** use the cosine rule or the sine rule for **right-angled triangles**.

H

Questions to try

1 The sides of a triangle are 1.5 cm, 2.5 cm and 3.5 cm respectively. Show that the triangle is an obtuse-angled triangle and find the size of the largest angle.

2 The diagram shows triangle ABC.

AB = 8.6 cm, BC = 3.1 cm and AC = 9.7 cm.

Calculate the angle ABC.

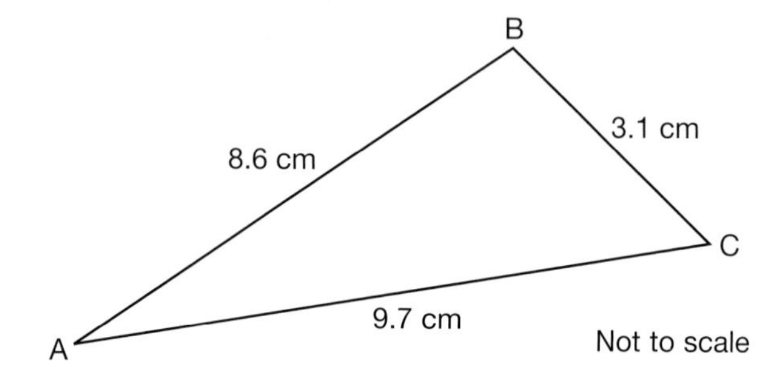

3 A ship leaves port and sails on a bearing of 032° for 12 km. The ship then changes direction and sets sail on a bearing of 058° for 15 km.

What is the distance and bearing of the ship from the port?

4 A, B and C lie in a straight line on horizontal ground.

BD is a vertical tower.
The angle of elevation of D from A is 38°.
The angle of elevation of D from C is 46°.
AC = 1000 m.

Calculate the height of the tower.

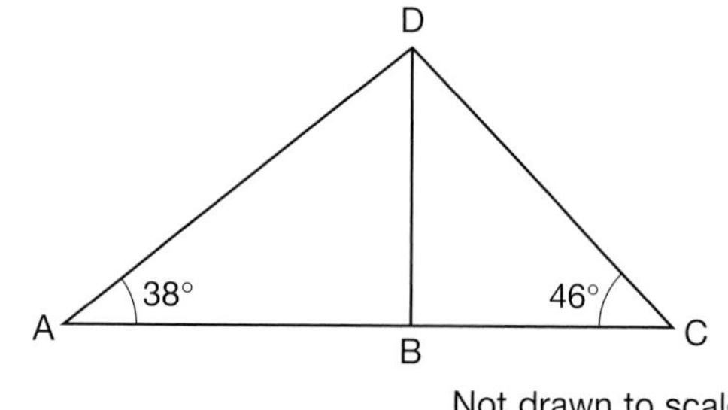

5 In triangle ABC the length of AB is 8.3 cm and angle ABC is 20°.

D is a point on BC such that the length of DC is 6.1 cm and angle ADB is 105°.

Calculate the length of AC.

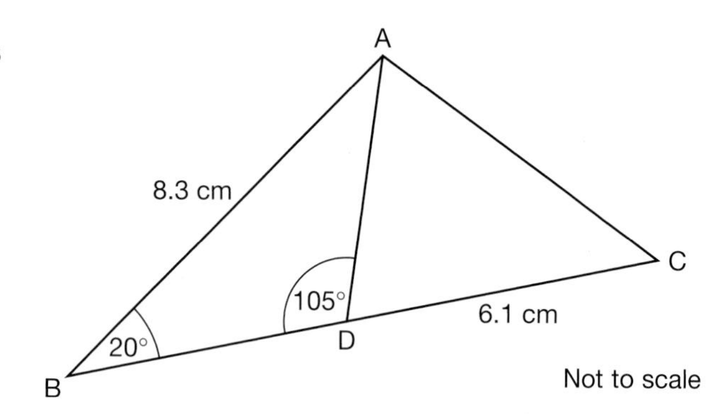

The answers can be found on pages **99–101.**

Exam Questions and Answers H

1 The diagram shows a trapezium ABCD.
AB is parallel to DC and is twice its length.

Given that **AB** = **p** and **DA** = **q**, express each of these vectors in terms of **p** and **q**.

(a) **AC** (b) **CB**

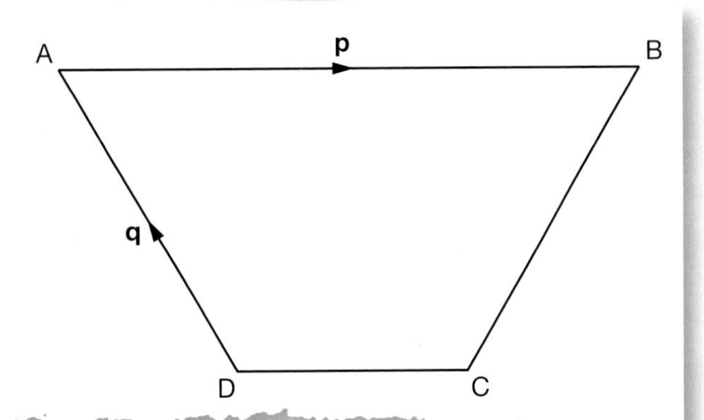

Answer

(a) $\overrightarrow{AC} = \frac{1}{2}p - q$

(b) $\overrightarrow{CB} = \frac{1}{2}p + q$

How to score full marks

- $\mathbf{AC} = \mathbf{AD} + \mathbf{DC}$

  $= -\mathbf{q} + \frac{1}{2}\mathbf{AB}$ (as AB is twice the length of DC)

  $= -\mathbf{q} + \frac{1}{2}\mathbf{p}$

  $= \frac{1}{2}\mathbf{p} - \mathbf{q}$

- The last line is not essential, but it looks tidier.

- $\mathbf{CB} = \mathbf{CD} + \mathbf{DA} + \mathbf{AB}$

  $= -\frac{1}{2}\mathbf{p} + \mathbf{q} + \mathbf{p}$

  $= \frac{1}{2}\mathbf{p} + \mathbf{q}$

- This answer uses the fact that **CD** = **–DC.** If the letters of a vector are reversed, the vector is acting in the opposite direction. This can often be useful when solving questions like this.

2 In triangle AOB, $\overrightarrow{OA} = \mathbf{a}$ and $\overrightarrow{OB} = \mathbf{b}$.
X is a point on OA such that OX : XA = 3 : 1.
Y is a point on AB such that AY : YB = 3 : 1.

(a) Express:
  (i) $\overrightarrow{OX}$ in terms of **a**
  (ii) $\overrightarrow{AY}$ in terms of **a** and **b**
  (iii) $\overrightarrow{XY}$ in terms of **a** and **b**.

(b) The line XY is extended to a point P so that $YP = \frac{1}{2}XY$.
  (i) Express $\overrightarrow{BP}$ in terms of **a** and **b**.
  (ii) Explain why O, B and P are in a straight line.

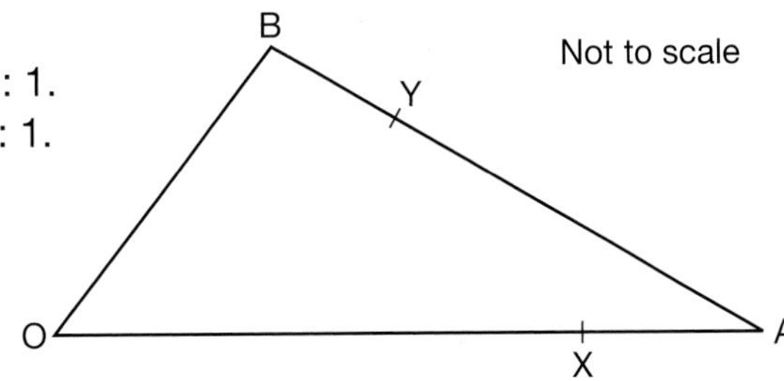

H

2

Answer

(a)(i) $\overrightarrow{OX} = \frac{3}{4}\underline{a}$

(ii) $\overrightarrow{AY} = \frac{3}{4}(\underline{b} - \underline{a})$

(iii) $\overrightarrow{XY} = \frac{3}{4}\underline{b} - \frac{1}{2}\underline{a}$

(b)(i) $\overrightarrow{BP} = \frac{1}{8}\underline{b}$

(ii) Since $\overrightarrow{OB} = \underline{b}$ and $\overrightarrow{BP} = \frac{1}{8}\underline{b}$ then the vectors are parallel. Therefore the lines OB and BP are parallel so that O, B and P must lie on a straight line.

How to score full marks

► $\overrightarrow{OX} = \frac{3}{4}\overrightarrow{OA}$ (the line OA is divided into 3 + 1 parts)

$= \frac{3}{4}\mathbf{a}$

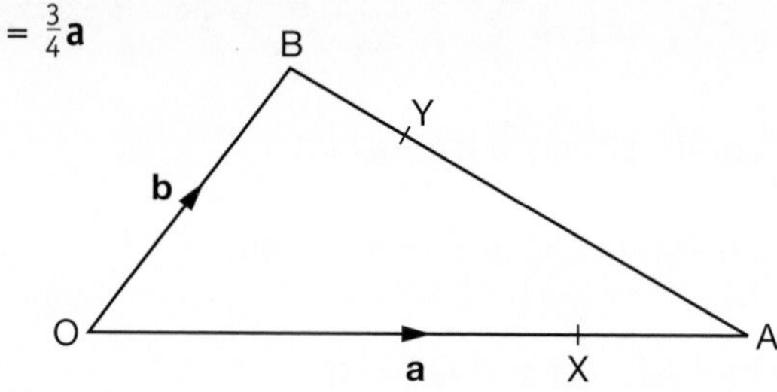

► $\overrightarrow{AY} = \frac{3}{4}\overrightarrow{AB}$ (the line AB is divided into 3 + 1 parts)

$= \frac{3}{4}(\overrightarrow{AO} + \overrightarrow{OB})$

$= \frac{3}{4}(-\mathbf{a} + \mathbf{b})$

$= \frac{3}{4}(\mathbf{b} - \mathbf{a})$

► $\overrightarrow{XY} = \overrightarrow{XA} + \overrightarrow{AY}$

$= \frac{1}{4}\mathbf{a} + \frac{3}{4}(\mathbf{b} - \mathbf{a})$ (using results from (a)(ii))

$= \frac{1}{4}\mathbf{a} + \frac{3}{4}\mathbf{b} - \frac{3}{4}\mathbf{a}$

$= \frac{3}{4}\mathbf{b} - \frac{1}{2}\mathbf{a}$ (tidying up the expression)

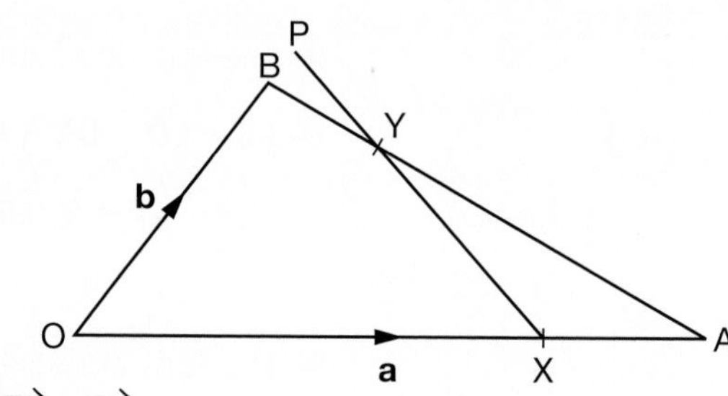

► $\overrightarrow{BP} = \overrightarrow{BY} + \overrightarrow{YP}$

$= \frac{1}{4}\overrightarrow{BA} + \frac{1}{2}\overrightarrow{XY}$

$= \frac{1}{4}(-\mathbf{b} + \mathbf{a}) + \frac{1}{2}(\frac{3}{4}\mathbf{b} - \frac{1}{2}\mathbf{a})$

$= -\frac{1}{4}\mathbf{b} + \frac{1}{4}\mathbf{a} + \frac{3}{8}\mathbf{b} - \frac{1}{4}\mathbf{a}$

$= \frac{1}{8}\mathbf{b}$

Key points to remember

Showing vectors

Vectors can be represented in a number of ways such as:

PQ or you can write $\overrightarrow{PQ}$

r or you can write $\underline{r}$

or as a column vector $\begin{pmatrix} s \\ t \end{pmatrix}$

This form is also used to represent translations, which are vectors since they act in a particular direction.

DON'T MAKE THESE MISTAKES ...

✗ Remember that the **direction of a vector** is important so always **show it with an arrow**.

✗ If a line is **divided in the ratio $x : y$** then there are **$x + y$ parts**.

The respective fractions are $\frac{x}{x+y}$ and $\frac{y}{x+y}$.

Equal vectors

Two vectors are equal if they have the same magnitude and direction – i.e. they are the same length and they are parallel.

Addition and subtraction of vectors

To add or subtract vectors, place them end to end with the arrows 'following on' in the same direction.

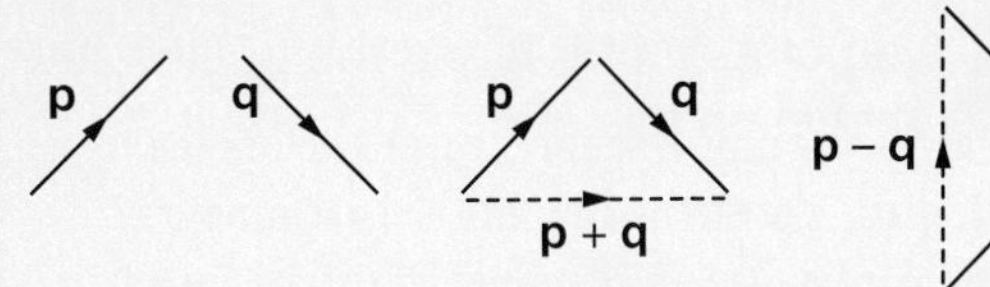

Components of a vector

The components of a vector are usually described in terms of:

(i) the number of units in the x-direction
(ii) the number of units in the y-direction

These units are best expressed as a column vector.

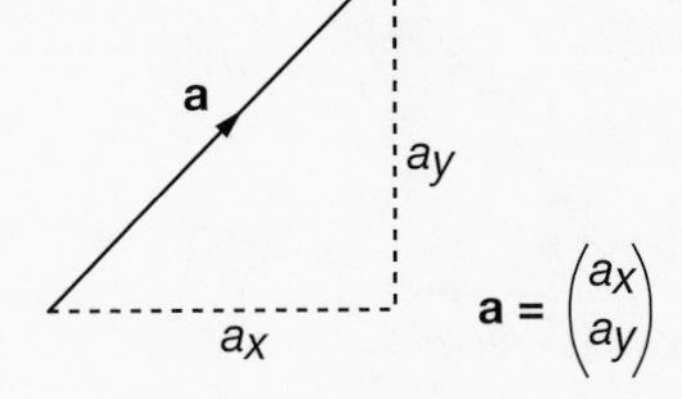

Multiplication of a vector

Vectors cannot be multiplied by other vectors but they can be multiplied by a constant (sometimes called scalar multiplication).

Magnitude of a vector

The magnitude (length) of a vector can be found by using Pythagoras' theorem.

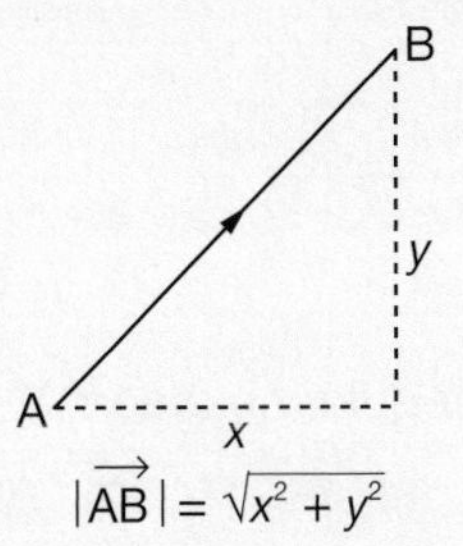

$|\overrightarrow{AB}| = \sqrt{x^2 + y^2}$

The length of the vector $\overrightarrow{AB} = \sqrt{x^2 + y^2}$.

You can write $|\overrightarrow{AB}| = \sqrt{x^2 + y^2}$ where the two vertical lines stand for 'magnitude' or length.

H

Questions to try

1 Given that $\mathbf{p} = \begin{pmatrix}5\\4\end{pmatrix}$ and $\mathbf{q} = \begin{pmatrix}2\\-3\end{pmatrix}$ estimate these.

(a) $\mathbf{p} + \mathbf{q}$ (b) $2\mathbf{p} - 3\mathbf{q}$ (c) $|\mathbf{p} + \mathbf{q}|$

2 PQRS is a trapezium with PQ parallel to SR.

PQRT is a parallelogram.

TR is twice the length of ST.

Given that **QR** = **a** and **PQ** = **b**, express each of the following in terms of **a** and **b**.

(a) **PT** (b) **PR**

(c) **RT** (d) **SR**

(e) **PS**

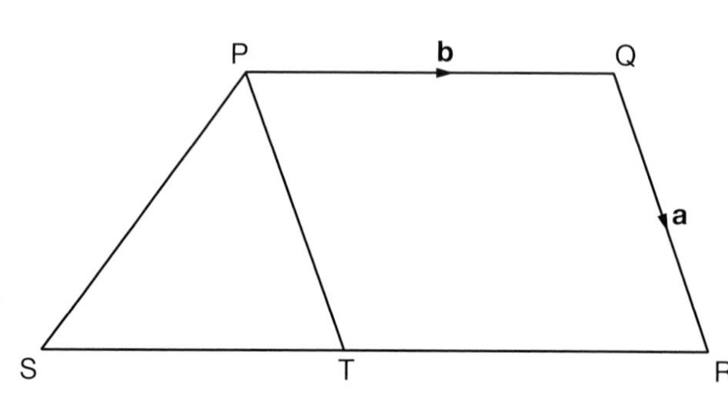

Diagram not drawn to scale

3 Vector **a**, vector **b** and the point X are shown on this grid.

(a) Vector $\overrightarrow{XY} = 3\mathbf{a} + 1\frac{1}{2}\mathbf{b}$ and vector $\overrightarrow{XZ} = 2\mathbf{a} - \frac{1}{2}\mathbf{b}$.

Mark Y and Z on the diagram.

(b) P is the midpoint of YZ.

Express $\overrightarrow{XP}$ in terms of **a** and **b**.

You **must** show all your working.

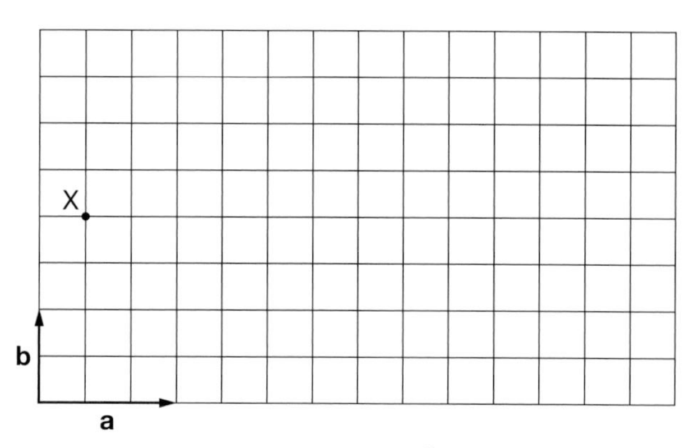

4 X and Y are the midpoints of PQ and PR respectively in the triangle PQR. Given that $\overrightarrow{XP} = \mathbf{a}$ and $\overrightarrow{PY} = \mathbf{b}$ show that QR = 2XY and QR is parallel to XY.

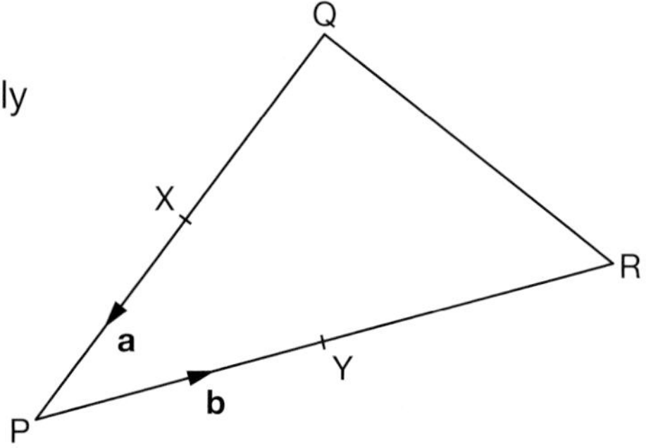

5 For the regular hexagon ABCDEF with centre O, $\overrightarrow{AB} = \mathbf{a}$ and $\overrightarrow{BC} = \mathbf{b}$.

Find: (a) $\overrightarrow{AC}$ (b) $\overrightarrow{AO}$ (c) $\overrightarrow{OB}$ (d) $\overrightarrow{AD}$.

What can you say about the quadrilateral ACDF?

Give reasons for your answer.

The answers can be found on pages **101–102.**

Exam Questions and Answers

1 The distance from the Earth to the sun is 9.3×10^7 miles.

Write this number in ordinary form.

Answer

$9.3 \times 10^7 = 9.3 \times 10 \times 10 \times 10 \times 10 \times 10 \times 10 \times 10$

$= 93\,000\,000$ miles

- Remember that $10^7 = 10 \times 10 \times 10 \times 10 \times 10 \times 10 \times 10$.
- Finally, remember to check that your answer is reasonable – and don't forget to include the units.

2 A rectangular picture measures 1.2×10^2 cm by 4.3×10^3 cm.

(a) What is the perimeter of the picture?

(b) What is the area of the picture?

Give your answers in standard form. Remember to state the units in your answers.

Answer

Perimeter

$= 1.2 \times 10^2 + 4.3 \times 10^3 + 1.2 \times 10^2 + 4.3 \times 10^3$

$= 120 + 4300 + 120 + 4300$

$= 8840$

$= 8.840 \times 10^3$ cm

Area

$= (1.2 \times 10^2) \times (4.3 \times 10^3)$

$= 1.2 \times 4.3 \times 10^2 \times 10^3$

$= 5.16 \times 10^{2+3}$

$= 5.16 \times 10^5$ cm^2

- Remember that the perimeter of a shape is the distance all around it.

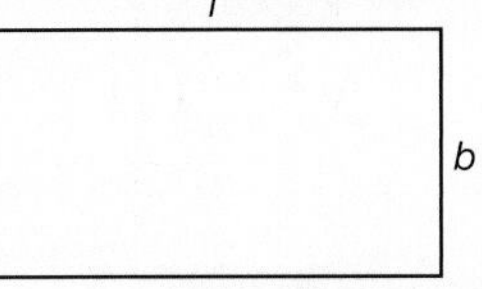

perimeter $= l + b + l + b$
$= 2l + 2b$

- When the power of 10 is the same in two numbers written in standard form, you can just add the decimal numbers together (i.e. $1.2 \times 10^2 + 1.2 \times 10^2 = 2.4 \times 10^2$ and $4.3 \times 10^3 + 4.3 \times 10^3 = 8.6 \times 10^3$) but otherwise you need to convert the numbers to ordinary form.
- The question says, 'Give your answers in standard form' so you will lose valuable marks if you don't.
- The question also says, 'Remember to state the units in your answers', so you must do this, too. It is good practice anyway!
- Remember that to find the area of a rectangle you multiply the length by the breadth (or width).

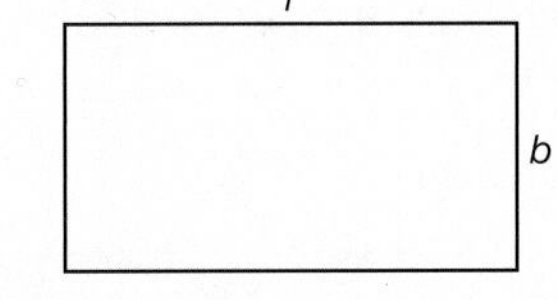

area $= l \times b$

- You should be able to multiply 1.2×4.3 without using a calculator (good practice for the non-calculator paper).
- You should be able to use the rules of indices to show that $10^2 \times 10^3 = 10^{2+3} = 10^5$.
- As the answer is already in standard form, you just need to include the units of area (cm^2) to gain full marks.

Key points to remember

You will need to recognise squares and cubes as well as work out square roots and cube roots.

Square numbers are made from other numbers that have been multiplied by themselves, e.g.

1, 4, 9, 16, 25, 36, …

Cube numbers are made from other numbers that have been multiplied by themselves then multiplied by themselves again, e.g.

1, 8, 27, 64, 125, 216, …

The square root of a number is the number that is squared to give that number, e.g. the square root of 36 is 6 (because $6 \times 6 = 36$).

The sign $^2\sqrt{}$ or $\sqrt{}$ is used to denote the square root, so $\sqrt{36} = 6$.

The cube root of a number such as 27 is the number that is cubed to give that number, e.g. the cube root of 27 is 3 (because $3 \times 3 \times 3 = 27$).

The sign $^3\sqrt{}$ is used to denote the cube root so $\sqrt[3]{27} = 3$.

You need to remember the rules of indices.

$a^m \times a^n = a^{m+n}$ so $3^7 \times 3^{12} = 3^{7+12} = 3^{19}$

$a^m \div a^n = a^{m-n}$ so $11^6 \div 11^2 = 11^{6-2} = 11^4$

$a^{-m} = \frac{1}{a^m}$ so $4^{-2} = \frac{1}{4^2} = \frac{1}{16}$

Don't forget them just because you are in an exam!

$a^0 = 1$

$a^1 = a$

Questions on standard form can appear on either the calculator or the non-calculator paper. On the calculator paper, you use the EXP or EE keys to input numbers in standard form.

When using your calculator, you must remember to interpret the display – of course, this will depend on your calculator.

3.62 07 means 3.62×10^7

4.01 -03 means 4.01×10^{-3}

DON'T MAKE THESE MISTAKES …

✗ **Don't multiply the number by the power.** Keep reminding yourself that:

$2^5 = 2 \times 2 \times 2 \times 2 \times 2 = 32$

$5^2 = 5 \times 5 = 25$

These are **wrong**:

$2^5 = 10$ ✗

$5^2 = 10$ ✗

Remember that 2^5 is bigger than 5^2.

Check that your answers are reasonable – remember that a positive power suggests a large number (e.g. $2.15 \times 10^6 = 2\,150\,000$) and a negative power suggests a small number (e.g. $2.001 \times 10^{-5} = 0.000\,020\,01$).

Questions to try

1

Work out 6×2^4.

2

Work out the value of:

(a) $(2^3)^2$

(b) 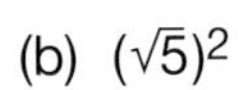 $(\sqrt{5})^2$

(c) $\sqrt{2^4 \times 25}$

3

Find, in standard form, the value of each of the following.

(a) $(3.42 \times 10^4) \times (5.91 \times 10^{-11})$

(b) 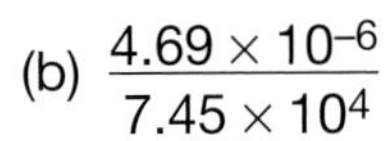 $\dfrac{4.69 \times 10^{-6}}{7.45 \times 10^4}$

4

If $p = 6 \times 10^{-2}$ and $q = 3 \times 10^{-4}$, calculate the value of:

(a) pq

(b) 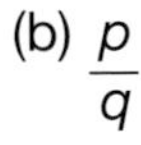 $\dfrac{p}{q}$

(c) $p - q$.

Give your answers in standard form.

5

Light travels at an average speed of 2.998×10^8 m/s.

The distance from the sun to the Earth is approximately 1.496×10^{11} m.

How long will it take light to travel from the sun to the Earth?

Give your answer to the nearest minute.

The answers can be found on page **103**.

8 Percentages

Exam Questions and Answers

1 **1** An investment valued at £3500 increases by 8% in one year. What is the new value of the investment?

1 **Answer**

1% of £3500 = £35
8% of £3500 = 8 × £35
= £280

New value = £3500 + £280
= £3780

How to score full marks

- To find 8% of £3500 you find:
  1% of £3500 = £35
  so 8% of £3500 = 8 × £35 = £280
- The new value of the investment is = £3500 + £280 = £3780
- Or use the fact that after an 8% increase, the new amount
  = 100% of the original amount + 8% of the original amount
  = 108% of the original amount
- The new value of the investment is 108% of £3500.
  1% of £3500 = £35
  108% of £3500 = 108 × £35 = £3780 (as before)

2 **2** The value of a picture is given as £127 200 after a 6% increase.
What was the original value of the picture?

2 **Answer**

106% of value = £127 200

1% of value = $\frac{£127\,200}{106}$
= £1200

100% of value = 100 × £1200
= £120 000

How to score full marks

- £127 200 is 106% of the original value (100% + 6%).
- So 106% of the original value = £127 200
  1% of the original value = $\frac{£127\,200}{106}$ (÷ 106)
  = £1200
- 100% of the original value = 100 × £1200 (× 100)
  = £120 000

Key points to remember

A big problem with percentage questions is that they can appear in so many different forms:

- express as a percentage
- find a percentage of a quantity
- find one number as a percentage of another
- calculate a percentage change
- use of percentages (e.g. simple or compound interest)
- reverse percentages

so make sure you know how to tackle each form.

Percentage change

To work out the percentage change, work out the increase or decrease first.

$$\text{percentage change} = \frac{\text{change}}{\text{original amount}} \times 100$$

where the change might be an increase, a decrease, a profit, a loss, an error, …

Use multipliers

If you are really sure you know what you are doing, you can use multipliers for percentage increase or percentage decrease questions, e.g.

- **an increase of 12% means**
  100% + 12% = 112% or $\frac{112}{100} = 1.12$
  If the price of an article costing £340 is increased by 12% then the new price is $1.12 \times £340 = £380.80$.
- **a decrease of 8% means**
  100% – 8% = 92% or $\frac{92}{100} = 0.92$
  If the price of an article costing £170 is decreased by 8% then the new price is $0.92 \times £170 = £156.40$.

DON'T MAKE THESE MISTAKES …

✗ **Don't just divide by 100** when you are asked to find a percentage – make sure you know how to do it.

✗ **Don't forget how to find one number as a percentage of another.**
For example, 30 as a percentage of 40 is $\frac{30}{40} \times 100 = 75\%$.

✗ **Don't be put off by reverse percentages**, where you are given the amount **after** the increase (or decrease). Always check your answer by working backwards.

VAT in your head

To find VAT you can use the fact that $17\frac{1}{2}\% = 10\% + 5\% + 2\frac{1}{2}\%$.

First, calculate 10%
(divide by 10 to find 10%).

Then calculate 5% (which is half of 10%).

Then calculate $2\frac{1}{2}\%$ (which is half of 5%).

Then add these amounts
($10\% + 5\% + 2\frac{1}{2}\% = 17\frac{1}{2}\%$)

which is easy (even without a calculator).

Questions to try

1 Find which of $\frac{5}{9}$, 0.7 and 63% is the least and which is the greatest.
You must show all your working.

2 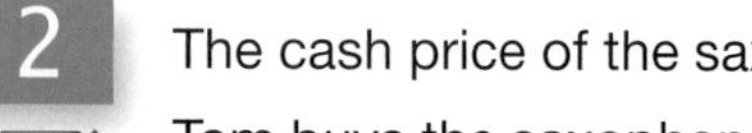

The cash price of the saxophone is £740.

Tom buys the saxophone using a credit plan.

He pays a deposit of 5% of the cash price and 12 monthly payments of £65.

Work out the difference between the cost when he used the credit plan and the cash price.

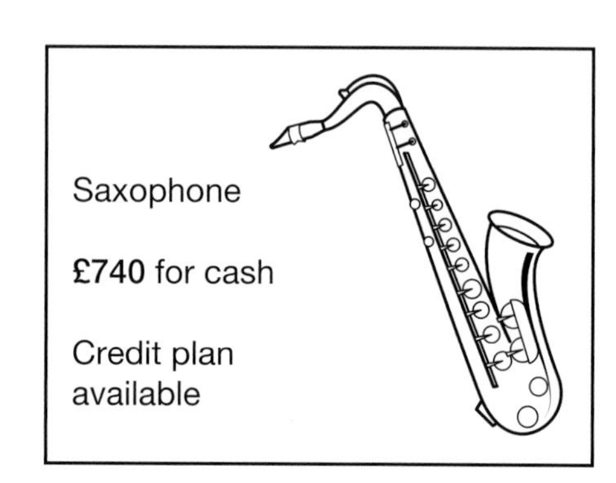

3 The population of a town increases by 14% each year. If the population is 35 000 this year then what will it be after:

(a) one year (b) two years?

4 Pratibha bought a car for £15 000.

Each year the value of the car depreciates by 10%.

Work out the value of the car two years after she bought it.

5 The weight of a child is recorded as 25 kg when his actual weight is 25.005 kg.

What is the percentage error on the actual weight?

6 A television is advertised at £550 after a price reduction of 12%.

What was the original price of the television?

 The answers can be found on page **104**.

Rational and irrational numbers 9

Exam Questions and Answers

1 (a) Which of the following numbers are rational?

$1 + \sqrt{2}$ $\quad \pi^2 \quad$ $3^0 + 3^{-1} + 3^{-2}$

(b) When p and q are two different irrational numbers, $p \times q$ can be rational.
Write down one example to show this.

Answer

(a) $1 + \sqrt{2}$ irrational

π^2 irrational

$3^0 + 3^{-1} + 3^{-2}$

$= 1 + \frac{1}{3} + \frac{1}{9}$

$= \frac{13}{9}$ so it is rational

(b) $\sqrt{2} \times \sqrt{8} = 4$

How to score full marks

- Remember that adding or subtracting a rational number and an irrational number always gives an irrational number.
- Remember that any multiple (or power) of π is always irrational.
- There are many examples to illustrate part (b):

The numbers must be different.

They must be irrational.

They must form the square root of a square number.

e.g.

$\sqrt{1} \times \sqrt{1} = 1$ ✗	$\sqrt{3} \times \sqrt{3} = 3$ ✗
$\sqrt{1} \times \sqrt{4} = 2$ ✗	$\sqrt{1} \times \sqrt{16} = 4$ ✗
$\sqrt{2} \times \sqrt{2} = 2$ ✗	$\sqrt{2} \times \sqrt{8} = 4$ ✓
$\sqrt{1} \times \sqrt{9} = 3$ ✗	$\sqrt{4} \times \sqrt{4} = 4$ ✗

You could also have
$\sqrt{3} \times \sqrt{12} = 6$
$\sqrt{5} \times \sqrt{20} = 10$
$\sqrt{6} \times \sqrt{24} = 12$ …
can you find some others?

2 $p = 2 + \sqrt{3} \qquad q = 2 - \sqrt{3}$

(a) (i) Work out $p - q$.

(ii) State whether $p - q$ is rational or irrational.

(b) (i) Work out pq.

(ii) State whether pq is rational or irrational.

Answer

(a) (i) $p - q = 2\sqrt{3}$

(ii) $p - q$ is irrational

(b) (i) $pq = 1$

(ii) pq is rational

How to score full marks

- $p - q = 2 + \sqrt{3} - (2 - \sqrt{3})$
  $= 2 + \sqrt{3} - 2 + \sqrt{3}$
  $= 2\sqrt{3}$

  Be careful with minus signs outside brackets.

- $2\sqrt{3}$ includes a surd, so it is irrational.

- $pq = (2 + \sqrt{3})(2 - \sqrt{3})$

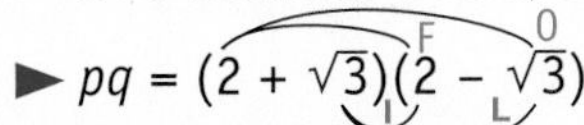

$= 4 - 2\sqrt{3} + 2\sqrt{3} - 3$ Remember FOIL.
$= 1$

H

Key points to remember

A rational number can be expressed in the form $\frac{p}{q}$ where p and q are integers.

Examples of rational numbers include $\frac{1}{5}$, $0.\dot{3}$, 7, $\sqrt{9}$, $\sqrt[3]{64}$.

Examples of irrational numbers include $\sqrt{2}$, $\sqrt{3}$, $\sqrt[3]{20}$, π.

Add some of your own examples here.

Rational	Irrational

Irrational numbers involving square roots are also called surds.

Surds can be multiplied and divided.

Remember the rules – there are only two, so it is quite simple!

$$\sqrt{a} \times \sqrt{b} = \sqrt{a \times b} \qquad \frac{\sqrt{a}}{\sqrt{b}} = \sqrt{\frac{a}{b}}$$

Try to recognise the following surds. They often come up on exam papers.

	1 ×	2 ×	3 ×	4 ×	5 ×	6 ×
All of these are multiples of $\sqrt{2}$	$\sqrt{2}$	$\sqrt{8}$	$\sqrt{18}$	$\sqrt{32}$	$\sqrt{50}$	$\sqrt{72}$
All of these are multiples of $\sqrt{3}$	$\sqrt{3}$	$\sqrt{12}$	$\sqrt{27}$	$\sqrt{48}$	$\sqrt{75}$	…
All of these are multiples of $\sqrt{5}$	$\sqrt{5}$	$\sqrt{20}$	$\sqrt{45}$	$\sqrt{80}$	…	
All of these are multiples of $\sqrt{7}$	$\sqrt{7}$	$\sqrt{28}$	$\sqrt{63}$	…		

DON'T MAKE THESE MISTAKES …

✗ **Don't be misled:**
$0.\dot{3}$ looks irrational but $0.\dot{3} = \frac{1}{3}$ so it is rational.

✗ **Don't be put off by repeated numbers.**
All recurring decimals are rational – take my word for it (or ask your teacher to show you why).

Questions to try

1 (a) Write down a rational number x such that $\sqrt{42} < x < \sqrt{44}$.

(b) Write down an irrational number y such that $6 < y < 7$.

(c) Write down a rational number for which the decimal form is never ending.

2 Simplify the following.

(a) $\sqrt{98}$

(b) $\sqrt{5} + \sqrt{45}$

(c) $\sqrt{6} \times \sqrt{12}$

3 Show that $(1 + \sqrt{2})(1 - \sqrt{2})$ is a rational number.

4 (a) $(3 - \sqrt{5}) \times (4 + \sqrt{5}) = a - \sqrt{5}$

Calculate the value of a.

(b) $(3 - \sqrt{5}) \times (4 + \sqrt{5}) + b$ is a rational number.

Give a possible value of b.

5 Given that $p = \sqrt{5}$, $q = \sqrt{10}$ and $r = \sqrt{8}$, simplify each of the following, indicating in each case whether your answer is rational or irrational.

(a) $pq - 1$

(b) $\dfrac{q}{pr}$

(c) $(p + q)^2$

6 Expand and simplify $(\sqrt{2} + \sqrt{3})^2$.

The answers can be found on page **105**.

10 Recurring decimals

Exam Questions and Answers

1

1 Convert $0.1\dot{5}$ to a fraction.

1

Answer

Let $0.1\dot{5} = x$.

Then $10x = 1.5555555 \ldots$

and $1x = 0.1555555 \ldots$

so $9x = 1.4$

$x = \frac{1.4}{9} = \frac{14}{90} = \frac{7}{45}$

How to score full marks

- This type of question is easy when you know how to do it. Keep practising these questions and you will soon get the hang of it.
- Notice that:
  $10 \times$ (fraction) $= 1.5555555\ldots$ (multiplying by 10)
  and $1 \times$ (fraction) $= 0.1555555\ldots$
- So subtracting gives:
  $9 \times$ (fraction) $= 1.4$ $(1.5555555\ldots - 0.15555555\ldots)$
  and (fraction) $= \frac{1.4}{9}$ (dividing both sides by 9)

  but $\frac{1.4}{9}$ is not a proper fraction, so using equivalent fractions:

  $\frac{1.4}{9} = \frac{14}{90} = \frac{7}{45}$ (cancelling $\frac{14}{90}$ to its lowest form)

  so $0.1\dot{5} = \frac{7}{45}$
- You can check by putting $\frac{7}{45}$ $(7 \div 45)$ into your calculator.

2

2 Convert $5.1\dot{5}$ to a fraction.

2

Answer

$5.1\dot{5} = 5 + 0.1\dot{5}$

Let $0.1\dot{5} = x$.

Then $10x = 1.555 \ldots$

and $1x = 0.1555 \ldots$

So $9x = 1.4$

$x = \frac{1.4}{9} = \frac{14}{90} = \frac{7}{45}$

So $5.1\dot{5} = 5 + \frac{7}{45}$

$= 5\frac{7}{45}$

How to score full marks

- Use the same method as above to convert $0.1\dot{5}$ to a fraction, then add the integer (whole number) part to it.
- $5.1\dot{5} = 5 + 0.1\dot{5} = 5\frac{7}{45}$

Key points to remember

Common recurring decimals

$\frac{1}{3} = 0.333333333333\ldots$ you write $0.\dot{3}$

$\frac{1}{6} = 0.166666666666\ldots$ you write $0.1\dot{6}$

$\frac{1}{7} = 0.142857142857\ldots$ you write $0.\dot{1}4285\dot{7}$

$\frac{1}{9} = 0.111111111111\ldots$ you write $0.\dot{1}$

$\frac{1}{11} = 0.090909090909\ldots$ you write $0.\dot{0}\dot{9}$

$\frac{1}{12} = 0.083333333333\ldots$ you write $0.08\dot{3}$

Note: All recurring decimals are rational numbers and can be converted to fractions.

Multiplying by 10, 100, 1000, . . .

Change $0.\dot{7}$ to a fraction.

Notice that $10 \times (\text{fraction}) = 7.77777777777\ldots$ $(\times 10)$

and $1 \times (\text{fraction}) = 0.77777777777\ldots$

Subtracting: $9 \times (\text{fraction}) = 7$

and $(\text{fraction}) = \frac{7}{9}$ (dividing both sides by 9)

so $0.\dot{7} = \frac{7}{9}$

Change $0.\dot{5}\dot{6}$ to a fraction.

Notice that: $100 \times (\text{fraction}) = 56.5656565656\ldots$ $(\times 100)$

and $1 \times (\text{fraction}) = 0.5656565656\ldots$

Subtracting $99 \times (\text{fraction}) = 56$

and $(\text{fraction}) = \frac{56}{99}$ (dividing both sides by 99)

so $0.\dot{5}\dot{6} = \frac{56}{99}$

Change $0.\dot{4}1\dot{3}$ to a fraction.

Notice that: $1000 \times (\text{fraction}) = 413.413413413\ldots$ $(\times 1000)$

and $1 \times (\text{fraction}) = 0.413413413\ldots$

Subtracting $999 \times (\text{fraction}) = 413$

and $(\text{fraction}) = \frac{413}{999}$ (dividing both sides by 999)

so $0.\dot{4}1\dot{3} = \frac{413}{999}$

In general terms:

$0.\dot{a} = \frac{a}{9}$

$0.\dot{a}\dot{b} = \frac{ab}{99}$

$0.\dot{a}b\dot{c} = \frac{abc}{999}$

where a, b and c are digits of the fraction.

If you remember this, then all you have to do is remember to cancel the fraction to its lowest terms.

Other recurring decimals

Look carefully to see which digits are repeating, then use the same sort of method as above.

DON'T MAKE THESE MISTAKES ...

✗ Be careful where you put those dots.

$0.1\dot{5} = 0.15555\ldots$

$0.\dot{1}\dot{5} = 0.151515\ldots$

✗ **Sometimes a whole group of digits is repeated.**

$\frac{1}{7} = 0.\dot{1}4285\dot{7}$

H

Questions to try

1

(a) Write the number $0.\dot{4}$ as a fraction in its simplest form.

(b) Write the number $0.\dot{5}$ as a fraction in its simplest form.

How do you think $0.\dot{7}$ and $0.\dot{8}$ can be written as fractions?

2

Write the number $0.\dot{2}\dot{1}$ as a fraction in its simplest form.

3

Convert $201.\dot{3}\dot{2}$ to a mixed number.

4

(a) Express $0.\dot{4}\dot{8}$ as a fraction in its simplest form.

(b) Hence, or otherwise, express $0.7\dot{4}\dot{8}$ as a fraction.

Write this fraction in its simplest form.

5

By considering $\frac{1}{9}$ as a decimal, write $0.7\dot{1}$ as a fraction in its lowest form.

The answers can be found on page **106.**

Exam Questions and Answers

1 A rectangle measures 12 cm by 7 cm, where each measurement is given to the nearest cm. Write down an interval approximation for the area of the rectangle.

Answer

Minimum area = 11.5 × 6.5
= 74.75 cm^2

Maximum area = 12.5 × 7.5
= 93.75 cm^2

Interval approximation
= 74.75 cm^2 to 93.75 cm^2

How to score full marks

- The minimum values for the two lengths are 11.5 cm and 6.5 cm (as the original lengths are given to the nearest cm).
- The lower bound (minimum area) is found by multiplying the minimum values of the two lengths.
- The maximum values for the two lengths are 12.5 cm and 7.5 cm (as the original lengths are given to the nearest cm).
- The upper bound (maximum area) is found by multiplying the maximum values of the two lengths.

2 A formula used in science is:

$$a = \frac{v - u}{t}$$

The quantities v, u and t are each measured correct to the nearest 0.1.

$u = 17.4$, $v = 30.3$ and $t = 2.6$.

Find the maximum possible value of a. You must show full details of your calculations.

Answer

Maximum value of

$$a = \frac{30.35 - 17.35}{2.55}$$

= 5.098 039 216

a = 5.1 (to 1 d.p.)

How to score full marks

- It will help if you work out the minimum and maximum values of u, v and t before going any further.

  $u_{min} = 17.35$ $v_{min} = 30.25$ $t_{min} = 2.55$

  $u_{max} = 17.45$ $v_{max} = 30.35$ $t_{max} = 2.65$

- Remember that to get the maximum value of the fraction $\frac{p}{t}$ you need to work out $\frac{p_{max}}{t_{min}}$.

  You need the maximum value of p (p_{max}) for the numerator (top line of the fraction), so work out $v_{max} - u_{min}$.

  You also need the minimum value of t (t_{min}) for the denominator (the bottom line of the fraction), which is 2.55.

- So, for the maximum value of a, work out:

  $$a = \frac{v_{max} - u_{min}}{t_{min}} = \frac{30.35 - 17.35}{2.55}$$

  to give a = 5.098 039 216 or a = 5.1 to 1 d.p.

- It is best to round your answer to 1 d.p. as the numbers given in the question are given correct to the nearest 0.1.

H

Key points to remember

Rounding . . .

- Most measures are rounded. For questions on this topic you need to know how to round:
  - to the nearest thousand
  - to the nearest whole number
  - to the nearest 0.1… .

. . . and boundaries

If a number is rounded to the nearest thousand, then the boundaries (upper and lower limits) are given by:

number $\pm \frac{1000}{2}$ (i.e. 'number – 500' to 'number + 500')

So if 5000 is given to the nearest 1000 then the boundaries are given by:

$5000 \pm \frac{1000}{2}$

which gives a lower bound of 5000 – 500 or 4500,
and an upper bound of 5000 + 500 or 5500.

Similarly a number rounded to the nearest whole number will have boundaries given by:

number $\pm \frac{1}{2}$ (i.e. 'number – $\frac{1}{2}$' to 'number + $\frac{1}{2}$')

and a number rounded to the nearest 0.1 will have boundaries given by:

number $\pm \frac{0.1}{2}$ (i.e. 'number – $\frac{0.1}{2}$' to 'number + $\frac{0.1}{2}$').

Finally, you need to remember that:

To find the **maximum value of $p - q$** you need to work out $p_{max} - q_{min}$ and to find the **minimum value of $p - q$** you need to work out $p_{min} - q_{max}$.

To get the **maximum value of $p \div q$** you need to work out $p_{max} \div q_{min}$ and to get the **minimum value of $p \div q$** you need to work out $p_{min} \div q_{max}$.

Remember that common sense and constant checking are very good tools for these questions.

DON'T MAKE THESE MISTAKES …

- ✗ Don't apply this method to upper and lower bounds for **ages**.
- ✗ If a person's age is 21, it **doesn't** mean that it ranges from $20\frac{1}{2}$ to $21\frac{1}{2}$.
- ✗ **For an age of 21** the lower bound is 21. **For an age of 21** the upper bound is 22 (actually 21 years, 364 days, 23 hours, 59 minutes and 59 seconds, i.e. a split second before their 22nd birthday, except in a leap year…).

Questions to try

1 For each of the following write down the lower and upper bound.

(a) 5000 (to the nearest thousand)

(b) 5000 (to the nearest 100)

(c) 3.62 (correct to 2 decimal places)

(d) 240 (correct to 2 significant figures)

(e) $7\frac{1}{2}$ (to the nearest half)

2 The River Nile is 6695 km long, measured to the nearest kilometre.
What is the shortest possible length, in metres?

3 A box measures 12 cm by 9 cm by 10 cm, each to the nearest centimetre.
Find the lower bound and the upper bound for the volume of the box.

4 Cressida is 18 years old.
Write down:

(a) the minimum age that Cressida could be.

(b) the maximum age that Cressida could be.

5 Assume the Earth to be a sphere of diameter 12 750 km, correct to the nearest 50 km.

Land covers 29% of the Earth's surface, correct to the nearest whole number.

Calculate the smallest possible area of land on the Earth's surface.

Give your answer in standard form to an appropriate degree of accuracy.

6 A train travels 65 kilometres (to the nearest kilometre) in a time of 35 minutes (to the nearest minute). What are the maximum and minimum speeds of the train, in kilometres per hour? Give your answer to an appropriate degree of accuracy.

The answers can be found on page **107.**

12 Cumulative frequency

Exam Question and Answer

1 **1** The table shows the mathematics test results for 90 pupils.

The test was marked out of 60.

Mark (x)	**Frequency**	**Cumulative frequency**
$0 \leqslant x \leqslant 10$	12	12
$10 < x \leqslant 20$	15	
$20 < x \leqslant 30$	29	
$30 < x \leqslant 40$	16	
$40 < x \leqslant 50$	11	
$50 < x \leqslant 60$	7	

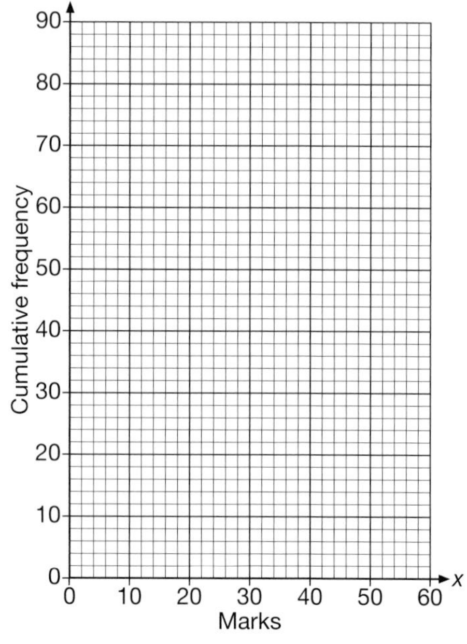

(a) Complete the cumulative frequency column.

(b) Draw a cumulative frequency diagram on the grid opposite.

(c) Use your diagram to estimate:

(i) the median

(ii) the inter-quartile range.

(d) The pass mark for the test is 25 marks.
Use your graph to estimate how many pupils pass the test.

1 **Answer**

(a)

Mark (x)	Frequency	Cumulative frequency
$0 \leqslant x \leqslant 10$	12	12
$10 < x \leqslant 20$	15	27
$20 < x \leqslant 30$	29	56
$30 < x \leqslant 40$	16	72
$40 < x \leqslant 50$	11	83
$50 < x \leqslant 60$	7	90

(b) See the graph opposite.

(c) (i) median = 26

(ii) IQR = 19

(d) 48

How to score full marks

- The cumulative frequency is a 'running total' of all the scores, so you need to add on the frequency each time.
- For the cumulative frequency diagram, plot the cumulative frequency against the upper value of the class.
- Read the values from the graph.
  - (i) the median (at the 45th value) = 26
  - (ii) the inter-quartile range (the difference between the upper quartile and the lower quartile) = 36 − 17 = 19
- From the graph, 42 pupils have fewer than 25 marks, so 90 − 42 = 48 pupils pass the test.

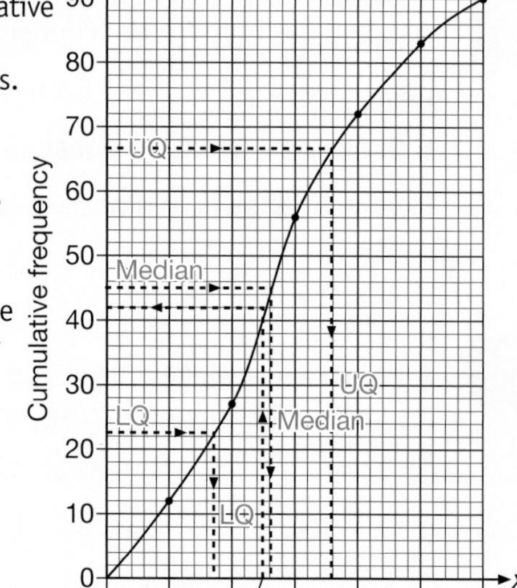

Key points to remember

The range

You can find the range of a distribution by working out the difference between the greatest value and least value.

Interquartile range

The interquartile range only takes account of the middle 50% of the distribution. You can find the interquartile range by dividing the data into four parts and working out the difference between the upper quartile and the lower quartile.

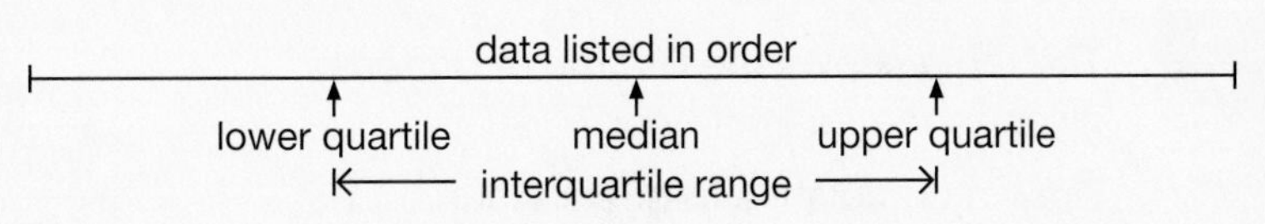

Finding the values

If there are n values in the distribution then:

- the median is the $\frac{1}{2}(n + 1)$th value
- the lower quartile is the $\frac{1}{4}(n + 1)$th value
- the upper quartile is the $\frac{3}{4}(n + 1)$th value

When n is large (say around 50) then:

- the median is the $\frac{1}{2}n$th value
- the lower quartile is the $\frac{1}{4}n$th value
- the upper quartile is the $\frac{3}{4}n$th value

Cumulative frequency diagrams

You can find the cumulative frequency by calculating the accumulated (or running) totals of the frequencies. Plot these against the upper class boundaries, then join up the points with a smooth curve to draw the cumulative frequency diagram (or ogive).

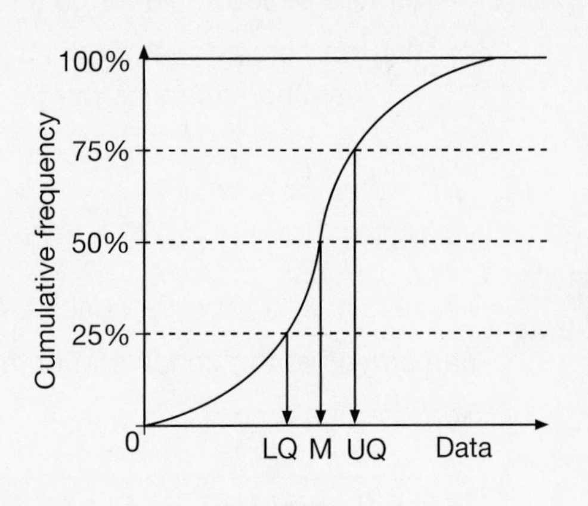

You must plot the points very carefully – the curve usually needs to be accurate to within $\pm \frac{1}{2}$ square.

DON'T MAKE THESE MISTAKES ...

- ✗ Don't lose easy marks – **always** present the **range** as a **single value** (for example, 7 not 11–18).
- ✗ **Always check** that the **accumulated** (or running) **totals add up** correctly.
- ✗ Remember to **plot the points** at the **upper class boundaries** – which is the highest value that the interval can take.
- ✗ All **cumulative frequency curves** have this shape.

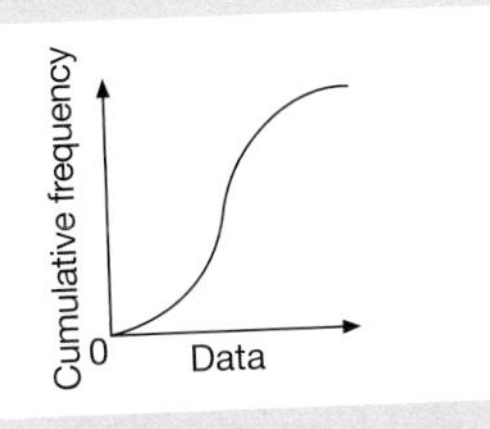

Questions to try

1 These are the heights of 15 bushes.

13 cm, 36 cm, 16 cm, 13 cm, 21 cm, 37 cm, 38 cm, 18 cm, 37 cm, 19 cm, 12 cm, 20 cm, 18 cm, 20 cm, 10 cm

Find the range and interquartile range.

2 The table shows the lifetime of batteries given to the nearest hour.

Draw a cumulative frequency curve to illustrate this information and use your graph to find:

(a) how many batteries lasted less than 250 hours

(b) how many batteries lasted more than 350 hours

(c) the median and the interquartile range.

Lifetime (hours)	Frequency
100–199	32
200–299	98
300–399	65
400–499	14
500–599	3
600–699	2

3 The speed in miles per hour (mph) of 200 cars travelling on a road were measured.

The results are shown in the table.

(a) Draw a cumulative frequency curve to show these figures.

(b) Use your graph to find an estimate for:

(i) the median speed (in mph)

(ii) the interquartile range (in mph)

(iii) the percentage of cars travelling at less than 48 miles per hour.

Speed (mph)	Cumulative frequency
not exceeding 20	1
not exceeding 25	5
not exceeding 30	14
not exceeding 35	28
not exceeding 40	66
not exceeding 45	113
not exceeding 50	164
not exceeding 55	196
not exceeding 60	200
Total	200

4 Michelle has to escort a coach party of 48 people.

She arrives at the coach station at 7 a.m.

She keeps a record of the time, t minutes, she has been waiting before people arrive.

Time (t) minutes	$5 \leqslant t < 10$	$10 \leqslant t < 15$	$15 \leqslant t < 20$	$20 \leqslant t < 25$
Frequency	0	2	4	7
Time (t) minutes	$25 \leqslant t < 30$	$30 \leqslant t < 35$	$35 \leqslant t < 40$	$40 \leqslant t < 45$
Frequency	15	14	5	1

(a) Complete the cumulative frequency table.

Time	$t < 10$							
Frequency	0							

(b) On the graph paper provided, draw a cumulative frequency diagram for the data.

(c) Find the median arrival time.

(d) Petra is the twelfth person to arrive.

 (i) Estimate her time of arrival.

 (ii) The coach arrived when three-quarters of the passengers were there. How long did Petra wait before the coach arrived?

(e) Karen arrived at 7:32 a.m. How many people arrived after Karen?

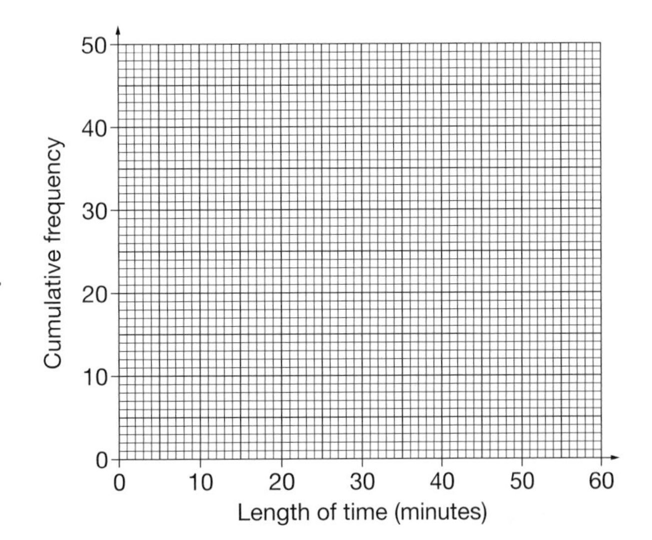

5 In a school survey, the heights of 80 boys aged 14 years are recorded.

The cumulative frequency graph shows the results.

(a) What is the median height of the boys?

(b) What is the interquartile range of the boys' heights?

The heights of 80 girls aged 14 are also recorded.

For the girls' heights:

 the median is 162 cm

 the lower quartile is 156 cm

 the upper quartile is 168 cm

 the shortest girl is 143 cm

 the tallest girl is 182 cm.

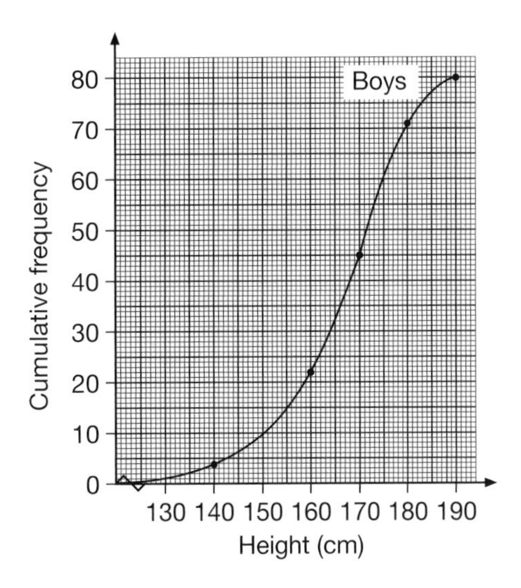

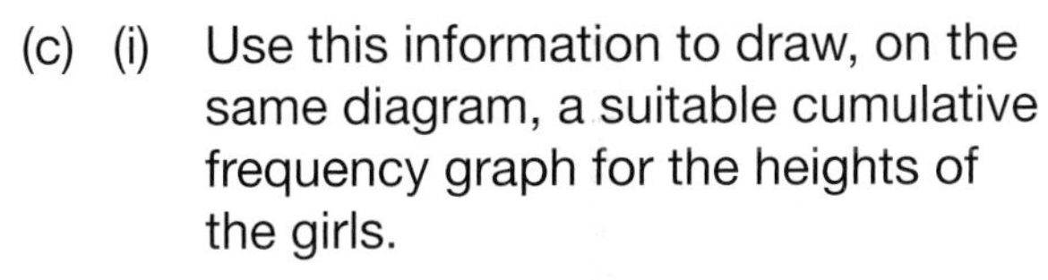

(c) (i) Use this information to draw, on the same diagram, a suitable cumulative frequency graph for the heights of the girls.

 (ii) Compare and comment on the heights of these boys and girls.

13 Mode, median and mean

Exam Questions and Answers

1

1 A student's test results are recorded as follows.

9 8 6 9 7 5 10 9 5 10

(a) Calculate:
(i) the mean mark
(ii) the median mark
(iii) the modal mark.

The student realises that one of the marks recorded as 5 should have been 6.

(b) What effect will this have on:
(i) the mean mark
(ii) the median mark
(iii) the modal mark?

1

Answer

(a) (i) Mean = 7.8

(ii) Median = $\frac{8 + 9}{2}$ = 8.5

(iii) Mode = 9

(b) (i) The mean mark will increase as the total mark increases slightly.

(ii) The median mark stays the same as the order is not affected, so the middle numbers are the same.

(iii) The modal mark stays the same as 9 is still the number that occurs most frequently.

How to score full marks

- The mean = $\frac{78}{10}$ = 7.8
- Arrange the numbers in order to find the median:

  5 5 6 7 8 9 9 9 10 10

  The median is the middle value = $\frac{8 + 9}{2}$ = 8.5

  The median is the $5\frac{1}{2}$th value, between the 5th and 6th.
- The mode is the value that appears most often and is 9.
- You need to explain briefly the effects you are describing but you do not need to recalculate.

2 The following table shows the numbers of words in 100 sentences of a particular book.

(a) Write down:

(i) the modal class interval

(ii) the class interval in which the median lies.

(b) Work out an estimate of the mean number of words in a sentence.

Number of words	Frequency f
1–5	15
6–10	29
11–15	25
16–20	15
21–25	11
26–30	2
31–35	1
36–40	0
41–45	2

Answer

(a) (i) Modal class interval = 6–10

(ii) Median falls in the 11–15 class.

(b) Estimated mean = 13

How to score full marks

► The modal class interval is the class with the greatest frequency. This is 6–10.

► The median is the $\frac{1}{2}(n + 1)$th value $= \frac{1}{2} \times 101 = 50\frac{1}{2}$

This falls in the 11–15 class.

Words	Frequency f	Mid-interval x	fx
1–5	15	3	45
6–10	29	8	232
11–15	25	13	325
16–20	15	18	270
21–25	11	23	253
26–30	2	28	56
31–35	1	33	33
36–40	0	38	0
41–45	2	43	86
	$\Sigma f = 100$		$\Sigma fx = 1300$

► Mean $= \frac{1300}{100} = 13$

Key points to remember

Mode

- The mode is the value that occurs most frequently.
- If there are two modes then the distribution is bimodal.
- If there are more than two modes then the distribution is multi-modal.

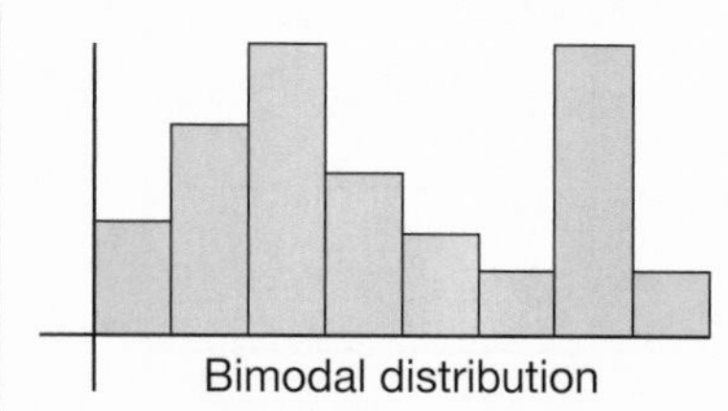
Bimodal distribution

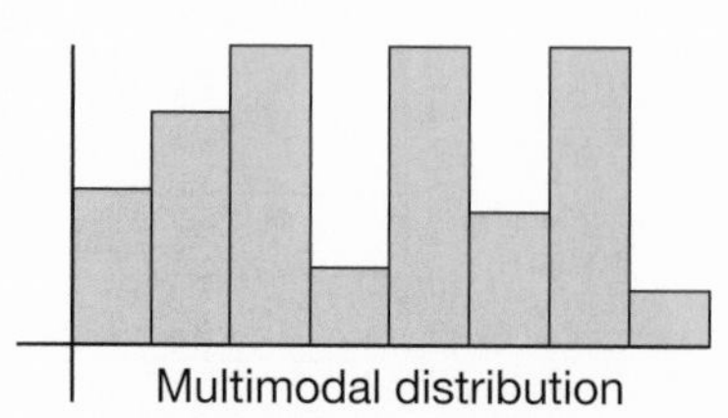
Multimodal distribution

Mode of a frequency distribution

- The mode of a frequency distribution is the value that has the highest frequency.
- The mode of a grouped frequency distribution is the group that has the highest frequency; this is called the modal group.

Median

- The median is the middle value when the values are arranged in order.
- Where there are two middle values, you add the two numbers and divide by 2.
- If there are n values in the distribution then the median position is given by the $\frac{1}{2}(n + 1)$th value.

Median of a frequency distribution

You can usually find the median of a frequency distribution or a grouped frequency distribution from a cumulative frequency distribution. See *Chapter 12, Cumulative frequency* for more information.

Mean

To find the mean of a distribution, find the total of the values of the distribution and divide by the number of values. The definition is often written as:

$$\text{mean} = \frac{\text{total of the values}}{\text{number of values}}$$

Mean of a frequency distribution

To find the mean of a frequency distribution, find the total of the values of the distribution (Σfx) and divide by the number of values (Σf).

$$\text{mean} = \frac{\Sigma fx}{\Sigma f}$$

Mean of a grouped frequency distribution

To find the mean of a grouped frequency distribution, use the mid-interval values or midpoints.

To find the mean of a grouped frequency distribution, find the total of the values of the frequency × mid-interval values of the distribution, then divide by the number of values.

$$\text{mean} = \frac{\text{total of (frequency} \times \text{mid-interval values)}}{\text{number of values}} = \frac{\Sigma fx}{\Sigma f}$$

DON'T MAKE THESE MISTAKES ...

✗ You could lose marks by giving silly answers. **Make sure your answers are realistic.**

✗ Don't guess at the **mid-interval value**. It is found by taking the **mean of the upper and lower class boundaries.**

✗ Don't forget that the **mid-interval values** are used as an **estimate** of the particular interval, so the **final answer** will not be exact but will be an **'estimate of the mean'**.

Questions to try

1 A set of 25 times, in seconds, is recorded.

12.9	10.0	4.2	16.0	5.6	18.1	8.3	14.0	11.5	21.7
22.2	6.0	13.6	3.1	11.5	10.8	15.7	3.7	9.4	8.0
6.4	17.0	7.3	12.8	13.5					

(a) Complete the frequency table below, using intervals of 5 seconds.

Time (t) seconds	Tally	Frequency
$0 \leqslant t < 5$		

(b) Write down the modal class interval.

2 David is playing cricket.

The table shows the number of runs he has scored off each ball so far.

Number of runs	0	1	2	3	4	5	6
Number of balls	3	8	4	3	5	0	2

(a) (i) What is the median number of runs per ball?

(ii) Calculate the mean number of runs per ball.

Off the next five balls, David scores the following numbers of runs.

4, 4, 5, 3 and 6

(b) (i) Calculate the new median.

(ii) Calculate the new mean.

(c) Give a reason why the mean is used, rather than the median, to give the average number of runs scored per ball.

3 (a) Pauline measures the lengths of some English cucumbers.

These are the lengths, in centimetres.

27, 28, 29, 30, 31, 31, 32, 33, 35, 37, 39

(i) What is the range of the lengths of these cucumbers?

(ii) What is the mean length of these cucumbers?

Pauline measures the lengths of some Spanish cucumbers. The range of the lengths of these cucumbers is 6 cm and the mean is 30 cm.

(b) Comment on the differences in these two varieties of cucumber.

4

The label on a jar of honey states that it contains 454 grams of honey.

The actual weight of honey in each of 20 jars was checked.

The results are shown in the table.

(a) Work out the range of the weights in the table.

(b) Calculate the mean weight of honey per jar.

Weight (g)	Number of jars
454	0
455	1
456	6
457	7
458	3
459	1
460	2

5

On holiday Val records the length of time people stay in a pool.

The results are shown in the table.

Calculate an estimate of the mean time spent in the pool.

Give your answer to an appropriate degree of accuracy.

Time, t (mins)	Number of people
$0 < t \leqslant 10$	4
$10 < t \leqslant 20$	7
$20 < t \leqslant 30$	3
$30 < t \leqslant 40$	2
Total	16

6

The following table shows the heights of plants.

Calculate an estimate of the mean height.

Height (cm)	15–20	20–30	30–40	40–50	50–60	60–70	70–80
Frequency	8	4	5	11	17	2	1

7

A manufacturer of school coats produces sizes according to height.

He measures the heights of 60 boys, with the following results.

Height, h (cm)	Frequency
$140 \leqslant h < 148$	6
$148 \leqslant h < 156$	10
$156 \leqslant h < 164$	15
$164 \leqslant h < 172$	18
$172 \leqslant h < 180$	11

(a) Calculate an estimate of the mean height of the boys.

(b) Explain how you can tell that the median lies between 156 cm and 164 cm.

(c) On a grid as shown, draw a histogram to display the data.

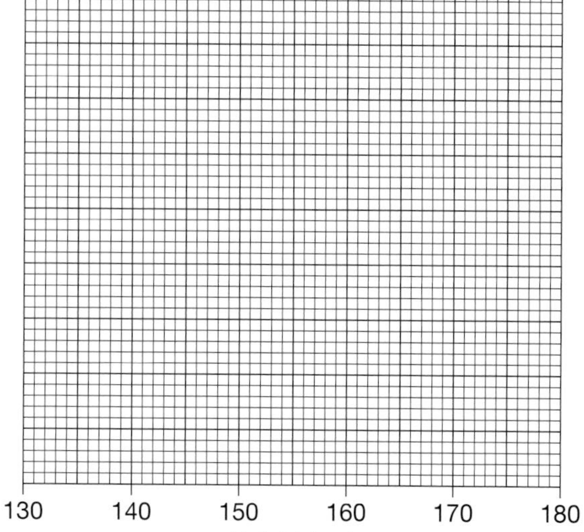

The answers can be found on pages **110–112.**

Exam Question and Answer

1 Nicky compared the scores awarded by two judges in a gymnastics competition.
Here are their scores for 10 competitors.

Judge 1	2.3	7.3	7.7	4.4	8.5	7.7	1.8	8.1	4.9	7.0
Judge 2	2.4	7.5	6.9	4.7	8.7	7.9	2.3	7.8	5.1	6.5

(a) Complete the scatter graph for these data. The first four points have been plotted for you.

(b) Describe fully the correlation between the two sets of scores.

(c) (i) Draw the line of best fit on the graph.

(ii) Estimate the score from Judge 2 for a competitor given 5.6 by Judge 1.

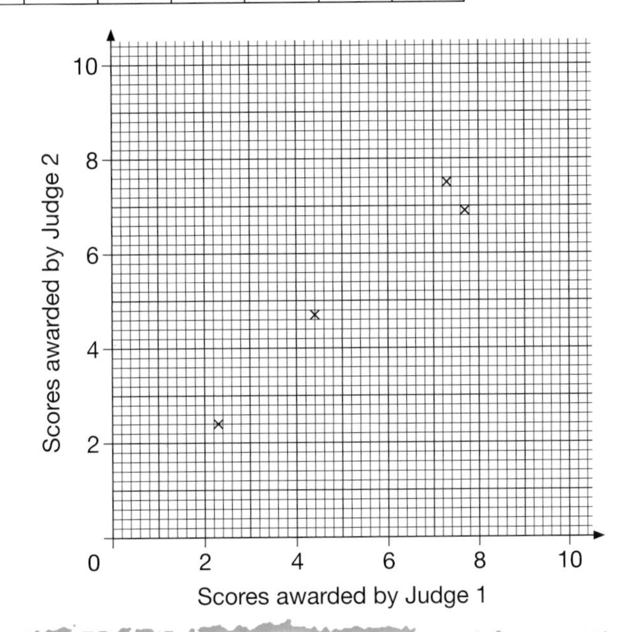

Answer

(a)

Scores awarded by Judge 2

Scores awarded by Judge 1

(b) *The correlation is strongly positive.*

(c) (i) *See the graph above.*

(ii) *Judge 2's score would be 5.7.*

How to score full marks

- Use the data to complete the scatter graph.
- Remember that you are asked to describe the correlation between the two sets of scores fully.
- Draw the line of best fit so that it passes through as many points as possible, with equal numbers of points on either side.
- Estimate the score from Judge 2 for a competitor given 5.6 by Judge 1.

Key points to remember

Scatter diagrams

Scatter diagrams (or scatter graphs) are graphs that are used to show the relationship between two variables. Each of the two variables is assigned to a different axis and the information is plotted as sets of coordinates.

Correlation

Scatter diagrams can be used to show whether there is any relationship or correlation between the two variables.

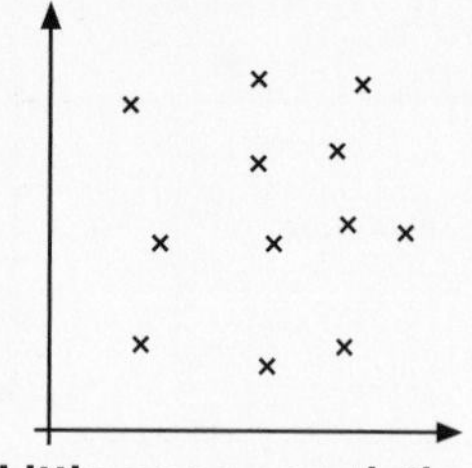

Little or no correlation

The points are scattered randomly over the graph, showing little or no correlation between the two variables.

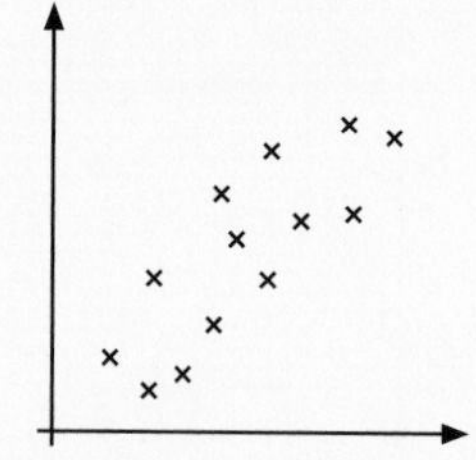

Moderate correlation

The points lie close to a straight line, suggesting some correlation between the two variables (the closer the points are to forming a line, the stronger the correlation).

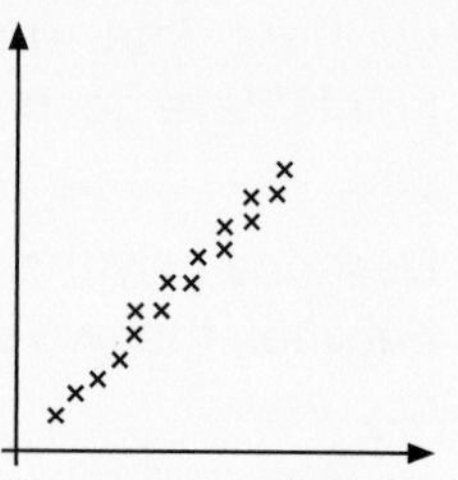

Strong correlation

The points lie almost along a straight line, showing a strong correlation between the two variables.

Correlation may also be positive or negative.

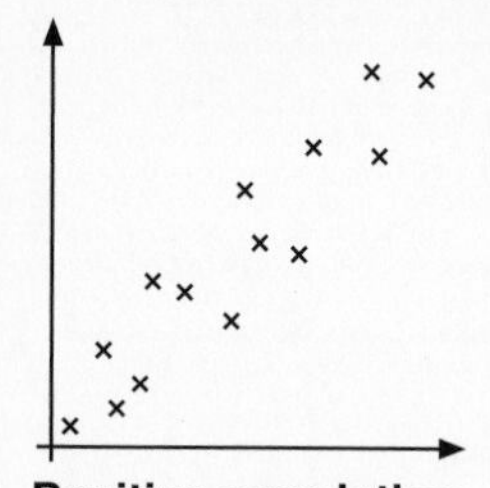

Positive correlation

Where an increase in one variable is associated with an increase in the other variable, the correlation is positive or direct.

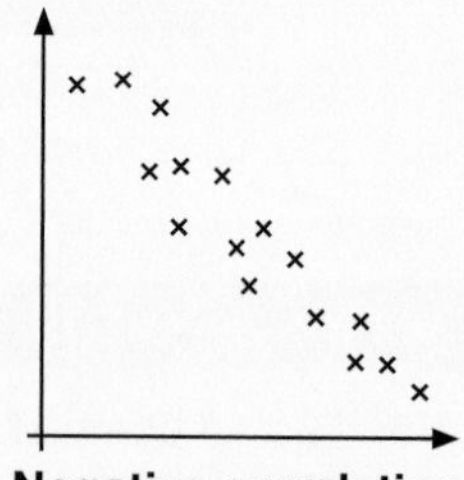

Negative correlation

Where an increase in one variable is associated with a decrease in the other variable, the correlation is negative or inverse.

Line of best fit

If the points on a scatter diagram show moderate or strong correlation, a line can be drawn to approximate the relationship.

This is called the line of best fit (or regression line).

This line can be used to predict other values from the given data.

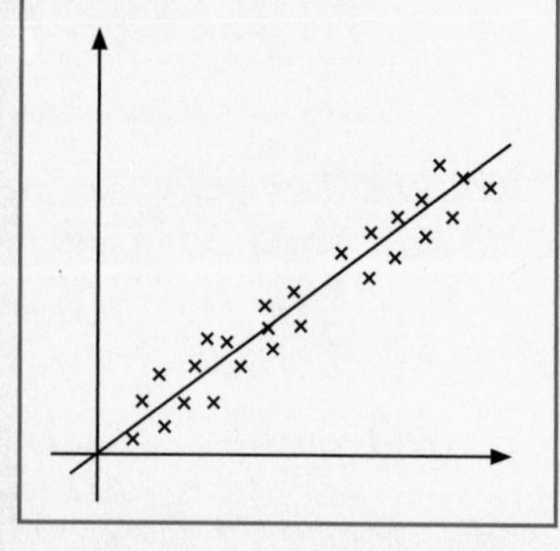

✗ Always draw the line of best fit so that there are **roughly equal numbers of points** on **either side** of the line.

✗ In most cases a **line of best fit** can be drawn 'by eye'.

✗ **The line of best fit does not have to pass through the origin**.

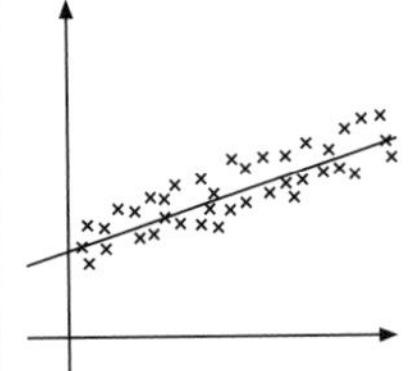

✗ For more accurate work, the **line of best fit should pass through** $(\bar{x}, \bar{y})$ where $\bar{x}$ and $\bar{y}$ are the **mean values of x and y** respectively.

Questions to try

1 The following table shows the heights and shoe sizes of 10 pupils.

Shoe size	3	2	5	$6\frac{1}{2}$	4	3	6	1	$3\frac{1}{2}$	$7\frac{1}{2}$
Height (cm)	132	125	149	157	134	127	151	117	141	100

Plot this information on a graph and comment on your findings.

2 The table shows the relationship between pupils' arm lengths and the distances they can throw an object.

(a) Plot this information on a graph and draw the line of best fit.

(b) Use the line of best fit to estimate:

(i) the distance thrown by someone with an arm length of 50 cm

(ii) the arm length when the distance an object is thrown is 5.5 m.

Arm length (cm)	Distance thrown (m)
40	4.9
36	4.2
33	3.0
46	5.7
43	4.8
41	4.6
47	6.0
52	6.1
45	5.4
33	3.6

3 The scatter diagram shows the height of some plants, d days after germinating.

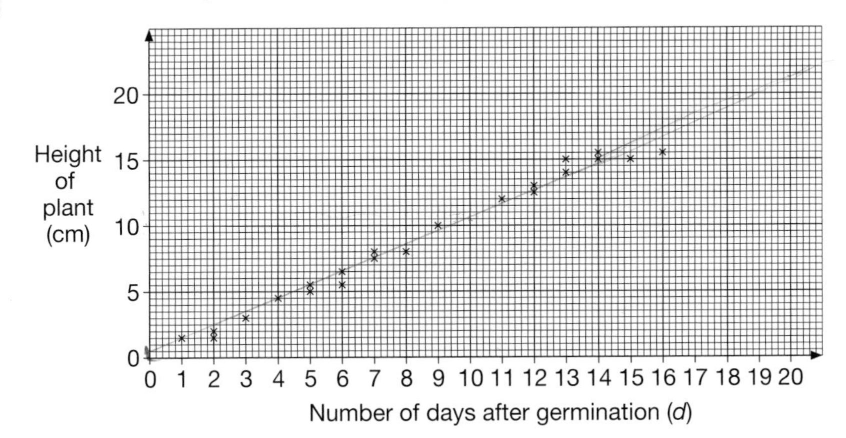

(a) Draw the line of best fit.

(b) Use your line of best fit to estimate the height of a plant:

(i) 10 days after germination

(ii) 20 days after germination.

(c) Which of your two answers in (b) is likely to be more reliable?

Give a reason for your answer.

4 The table shows the 1961 figures for adult literacy and life expectancy in eight countries.

Country	Adult literacy %	Life expectancy (years)
Chile	88	63
Ethiopia	10	38
India	29	49
Iran	49	51
Italy	90	71
Malaysia	58	58
Paraguay	75	61
Saudi Arabia	15	45

(a) On the graph paper, draw a scatter graph of these figures.

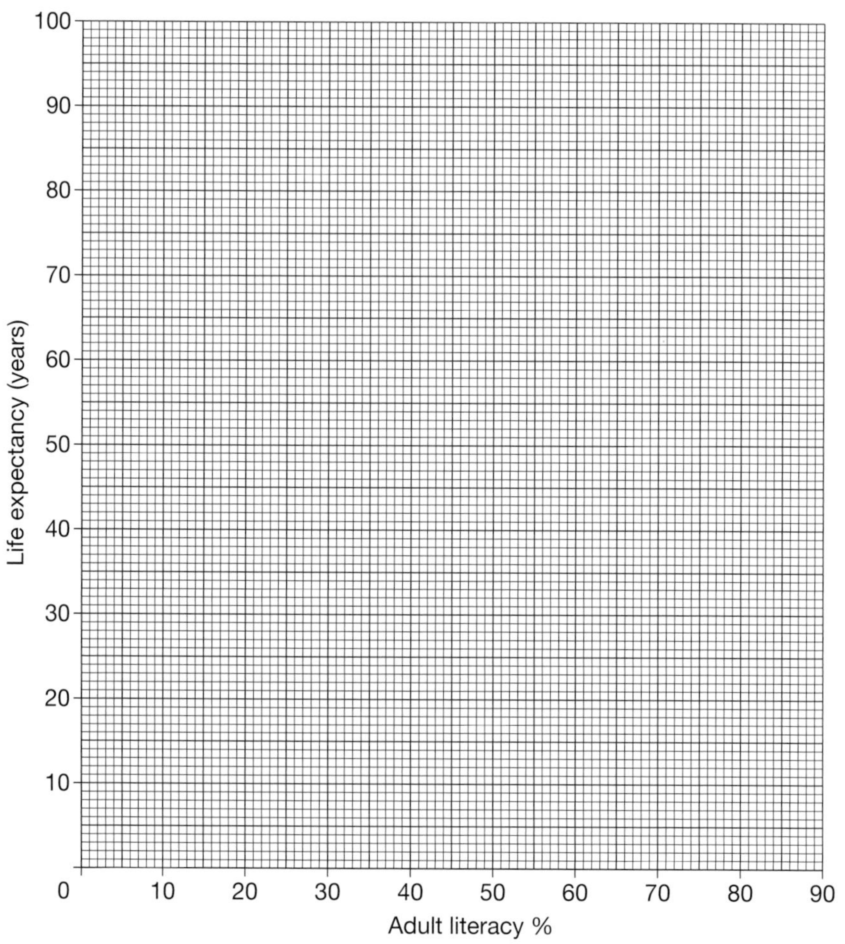

(b) From the scatter graph, state what correlation, if any, there is between 'Adult literacy' and the 'Life expectancy' figures.

(c) Draw a line of best fit.
Using this, estimate the life expectancy of an adult from a country with 42% adult literacy.

(d) Do you think that it is a good idea to draw conclusions from this set of figures? Give a reason for your answer.

 The answers can be found on pages **113–114.**

Exam Questions and Answers

1 Six events are given below.

Select a letter on the probability scale on the right which best shows the probability of these events happening.

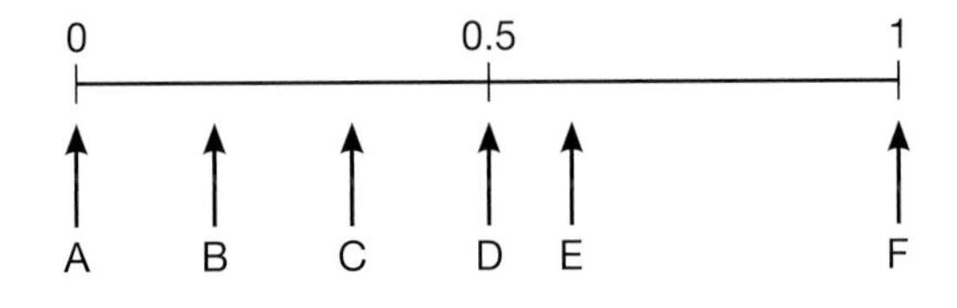

(i) throwing a six on a fair die
(ii) a newborn baby being a boy
(iii) selecting a month with 31 days from the calendar
(iv) February will have 31 days next year
(v) there will be 30 days in June
(vi) selecting a chocolate from a bag containing one mint, one chocolate and one jelly sweet

Answer

(i) B
(ii) D
(iii) E
(iv) A
(v) F
(vi) C

How to score full marks

- p(throwing a six) = $\frac{1}{6}$
- p(a newborn baby being a boy) = $\frac{1}{2}$
- p(selecting a month with 31 days from the calendar) = $\frac{7}{12}$
- p(February will have 31 days next year) = 0
- p(30 days in June) = 1
- p(selecting a chocolate) = $\frac{1}{3}$

2 A box contains buttons.

The buttons have two, three or four holes as shown.

The buttons are coloured white or red or green.

A button is taken at random from the box.

The table shows the probabilities of the number of holes and the colours of the buttons.

		Number of holes		
		2	3	4
Colour	White	0.3	0.2	0
	Red	0.1	0	0.2
	Green	0	0.1	0.1

(a) A button is taken from the box at random.
What is the probability that:
(i) it is red with two holes
(ii) it has two holes
(iii) it is red or has two holes (or both)
(iv) it has four holes?

(b) The box contains 50 buttons. How many buttons have four holes?

Answer

(a) (i) 0.1
(ii) 0.4
(iii) 0.6
(iv) 0.3
(b) 15

How to score full marks

- p(red with two holes) = 0.1 (from table)
- p(two holes) = 0.3 + 0.1 + 0 = 0.4
- p(red or two holes) = 0.1 + 0 + 0.2 + 0.3 + 0 (don't count 0.1 twice)
- p(four holes) = 0 + 0.2 + 0.1 = 0.3
- The number with four holes will be $50 \times 0.3 = 15$.

Key points to remember

Probability

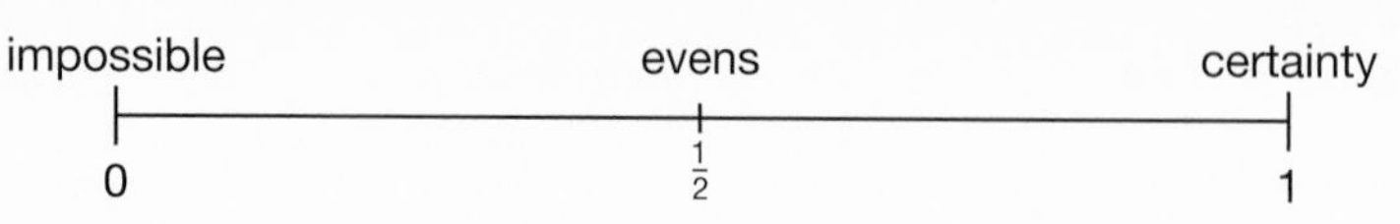

In probability:

- an event that is certain to happen has a probability of 1
- an event that is impossible has a probability of 0.

In general, $\text{p(success)} = \dfrac{\text{number of 'successful' outcomes}}{\text{number of 'possible' outcomes}}$

where p(success) means the probability of success.

Total probability

p(an event happening)
= 1 – p(an event not happening)

p(an event not happening)
= 1 – p(an event happening)

so p(an event happening)
+ p(an event not happening) = 1

Tree diagrams

In a tree diagram, the probabilities of events are written on different branches.

The total sum of the probabilities at the ends of all the branches must equal 1.

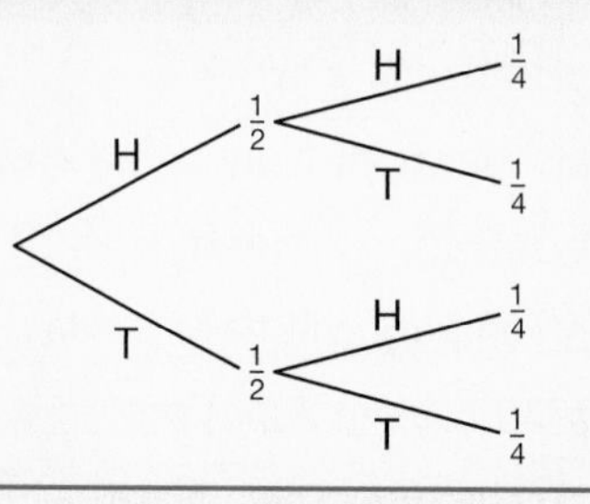

Possibility spaces

A possibility space is a diagram which can be used to show the outcomes of various events.

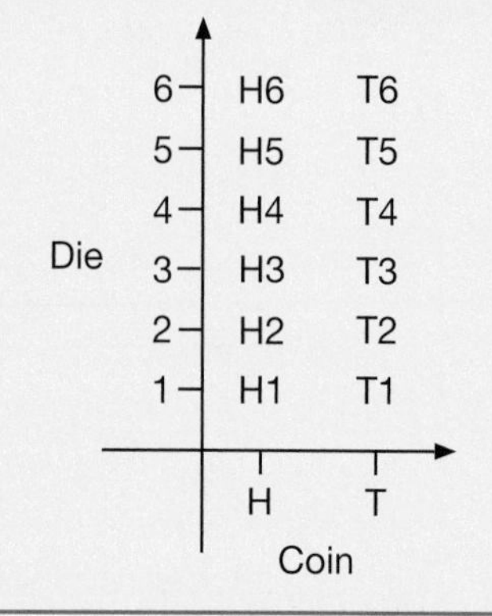

Addition rule for mutually exclusive events

Events are mutually exclusive if they cannot happen at the same time. For example, a playing card cannot be both a spade and a red card.

For mutually exclusive events, you can apply the addition rule (also called the **or** rule):

p(A or B) = p(A) + p(B)

Remember that to add fractions you must have a common denominator.

Multiplication rule for independent events

Events are independent if one event does not affect the other event. For example, if you toss a coin and pick a card, neither event affects the result of the other.

For independent events, you can use the multiplication rule (also called the **and** rule):

p(A and B) = p(A) × p(B)

Remember that to multiply fractions you multiply the numerators and you multiply the denominators.

DON'T MAKE THESE MISTAKES ...

- ✗ **Probabilities greater than 1 or less than 0** do not have any meaning.
- ✗ Probabilities can be expressed as **fractions, decimals or percentages but not as ratios.**

Questions to try

1 (a) A bag contains only red marbles and blue marbles.

Hugh chooses a marble without looking.

Under each of the statements, tick the appropriate box.

(i) The probability that the marble is red is $\frac{1}{2}$.

Must be correct ☐ Might be correct ☐ Must be wrong ☐

(ii) The probability that the marble is red or blue is 1.

Must be correct ☐ Might be correct ☐ Must be wrong ☐

(iii) The probability that the marble is red is –0.4.

Must be correct ☐ Might be correct ☐ Must be wrong ☐

(b) Another bag contains three red, four blue and five green marbles.
Hugh chooses a marble from this bag without looking.

What is the probability that it is:

(i) blue (ii) blue or red (iii) yellow?

2 The table shows the colour and make of 20 cars.

		Colour of car	
		White	**Blue**
Make of car	**Vauxhall**	7	4
	Ford	3	6

(a) A car is chosen at random.

What is the probability that it is a blue Ford?

(b) A white car is chosen at random.

What is the probability that it is a Vauxhall?

3 A record was kept of the number of occupants in cars arriving at a car park.

The numbers collected for 50 cars are shown in the bar chart.

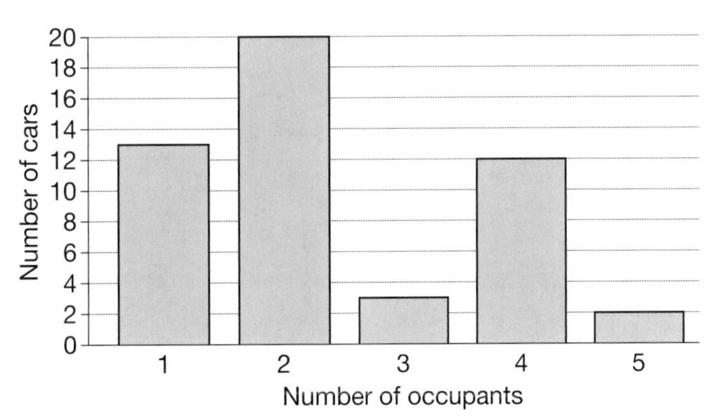

(a) One of these cars is chosen at random. Find the probability that it arrived with two occupants.

(b) Find the total number of car occupants.

(c) One of the car occupants is chosen at random.

Find the probability that the person was driving a car as it arrived at the car park.

4

(a) A fair coin is thrown 20 times. It lands heads 12 times.

What is the relative frequency of throwing a head?

The coin continues to be thrown. The table shows the number of heads recorded for 20, 40, 60, 80 and 100 throws.

Number of throws	20	40	60	80	100
Number of heads	12	18	30	42	49

(b) Draw a graph to show the relative frequency of throwing a head for these data.

(c) Estimate the relative frequency of throwing a head for 1000 throws.

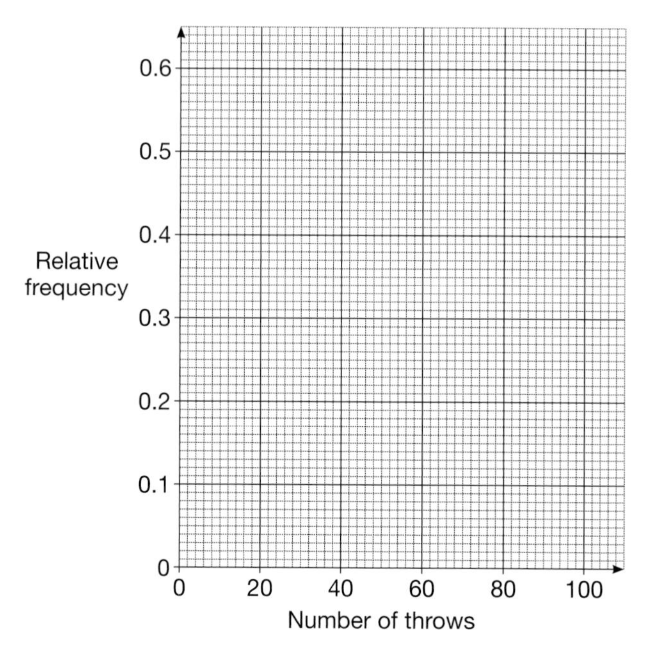

5

Andy and Baljit run in different races.

The probability that Andy wins his race is 0.8.

The probability that Baljit wins his race is 0.6.

(a) Fill in the missing probabilities on the tree diagram.

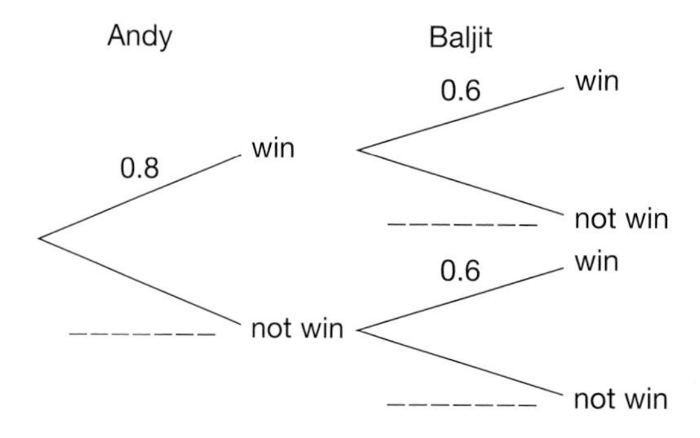

(b) Calculate the probability that only one of the boys wins his race.

6 In a game, Paula spins two fair five-sided spinners at the same time.
Her total score is the sum of the scores on the two spinners.

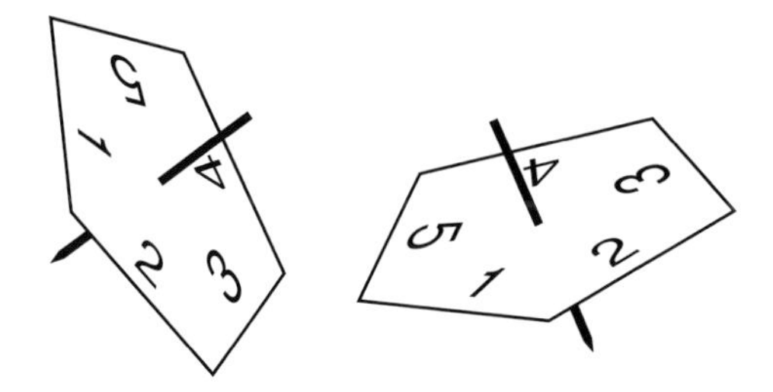

(a) What is the probability that Paula scores a total of 10 on her next spin?

(b) What is the probability that she scores a total of 9 on her next spin?

(c) What is the probability that the total score on her next spin is 3 or less?

(d) In another game, three of these spinners are used. What is the probability that all three of them show a five on the next spin?

7 Here is a five-sided spinner.

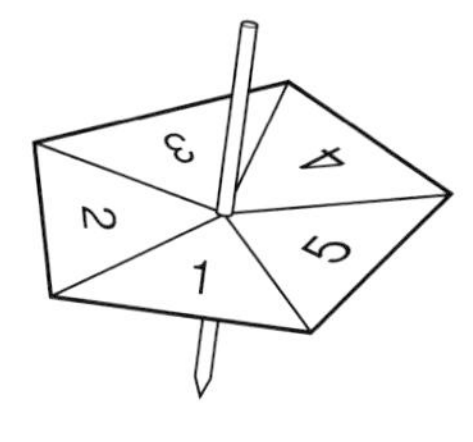

Its sides are labelled 1, 2, 3, 4, 5.

Alan spins the spinner and throws a coin.

One possible outcome is (3, Heads).

(a) List all the possible outcomes.

The spinner is biased.

The probability that the spinner will land on each of the numbers 1 to 4 is given in the table.

Number	1	2	3	4	5
Probability	0.36	0.1	0.25	0.15	

Alan spins the spinner once.

(b) (i) Work out the probability that the spinner will land on 5.

(ii) Write down the probability that the spinner will land on 6.

(iii) Write down the number that the spinner is most likely to land on.

(iv) Work out the probability that the spinner will land on an even number.

Alan spins the spinner and throws a fair coin.

(c) Work out the probability that the spinner will land on 3 and the coin will show Heads.

The answers can be found on pages **114–116.**

16 Histograms

Exam Question and Answer H

1

1 A farmer weighs all 60 eggs collected on Monday. The results are summarised in the grouped frequency distribution below.

Mass, x grams	Number of eggs, f	Frequency density
$30 \leqslant x < 50$	5	
$50 \leqslant x < 60$	15	
$60 \leqslant x < 90$	30	
$90 \leqslant x < 110$	10	

(a) Complete the frequency density column in the table and draw a histogram of the data in the table.

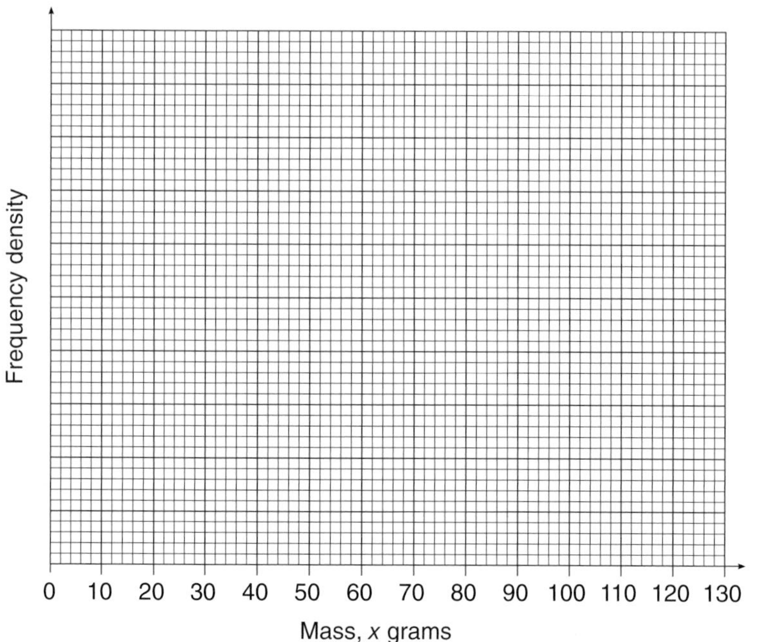

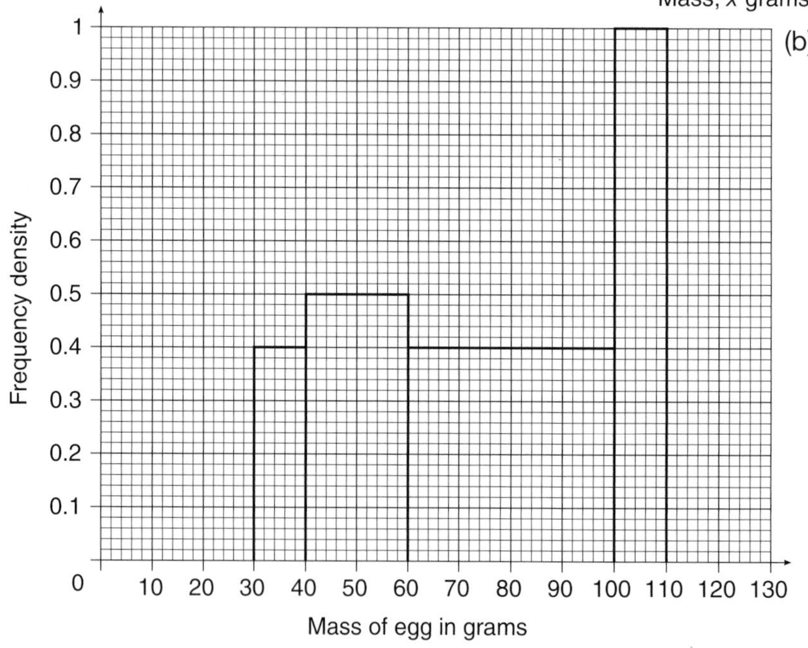

(b) The next day, Tuesday, the farmer again weighs all the eggs collected. The results are shown in the histogram on the left.

(i) Use this histogram to calculate how many eggs were collected on Tuesday.

(ii) Estimate the mass that is exceeded by 50% of the eggs on Tuesday.

Answer

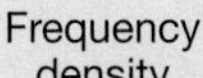

(a)

Mass, x grams	Number of eggs, f	Frequency density
$30 \leqslant x < 50$	5	0.25
$50 \leqslant x < 60$	15	1.5
$60 \leqslant x < 90$	30	1
$90 \leqslant x < 110$	10	0.5

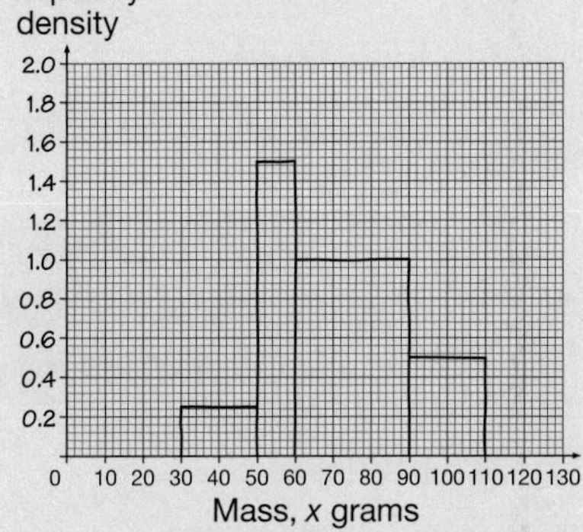

(b) (i) 40

(ii) 75 g

How to score full marks

► Complete the frequency density column in the table, by calculating (frequency ÷ class width) for each row.

Mass, x grams	Number of eggs, f	Class width	Frequency density
$30 \leqslant x < 50$	5	20	$5 \div 20 = 0.25$
$50 \leqslant x < 60$	15	10	$15 \div 10 = 1.5$
$60 \leqslant x < 90$	30	30	$30 \div 30 = 1$
$90 \leqslant x < 110$	10	20	$10 \div 20 = 0.5$

► Use the table to draw the histogram.

► (i) The number of eggs is the area of the bars in the histogram:
$= 10 \times 0.4 + 20 \times 0.5 + 40 \times 0.4 + 10 \times 1 = 40$ eggs

(ii) To estimate the mass that is exceeded by 50% of the eggs on Tuesday, you need to draw a line to represent 20 eggs.

The line needs to be drawn $\frac{6}{16}$ along the interval.

The value is $60 + \frac{6}{16} \times 40 = 60 + 15 = 75$ g

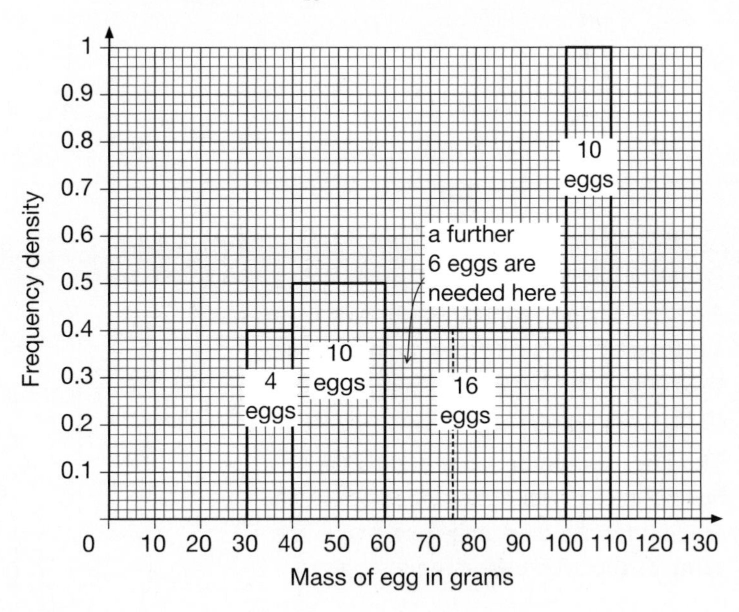

H

Key points to remember

Histograms

In histograms, the area of each bar represents the frequency so:

- class width × height = frequency
- height = $\frac{\text{frequency}}{\text{class width}}$

The height is the frequency density, so remember that the vertical axis of a histogram should be labelled frequency density and:

- frequency density = $\frac{\text{frequency}}{\text{class width}}$

For equal class intervals, the histogram looks like a bar chart.

DON'T MAKE THESE MISTAKES ...

✗ **Histograms look like bar charts but don't forget that it is the area of each 'bar' that represents the frequency, not the length or height.**

✗ The bars on the **horizontal axis** should be drawn at the **class boundaries for measured quantities such as heights and masses.**

Histograms and bar charts

They are alike if the bars are of equal width, but it is the **area** of each bar that represents the frequency.

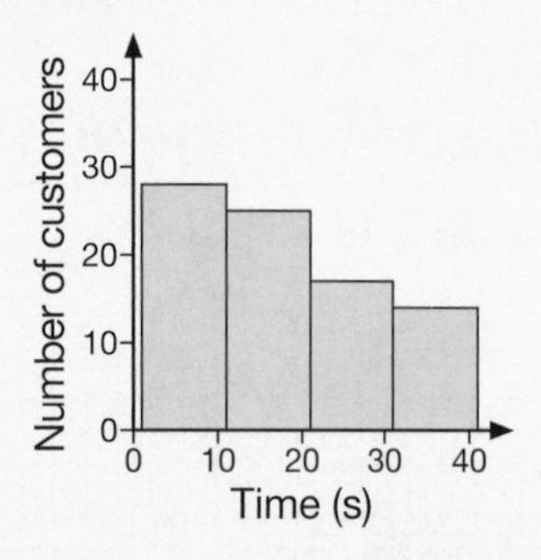

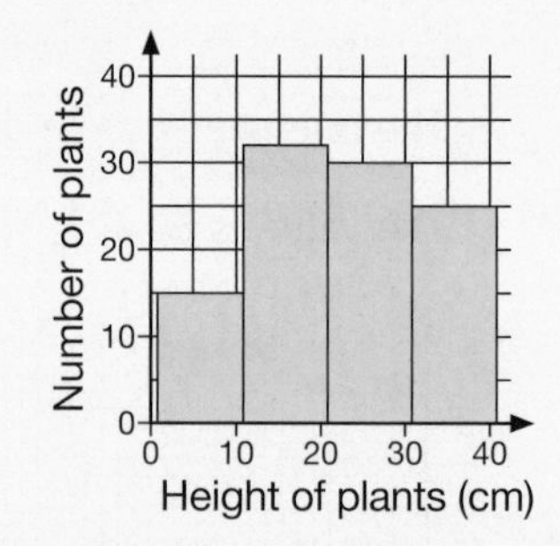

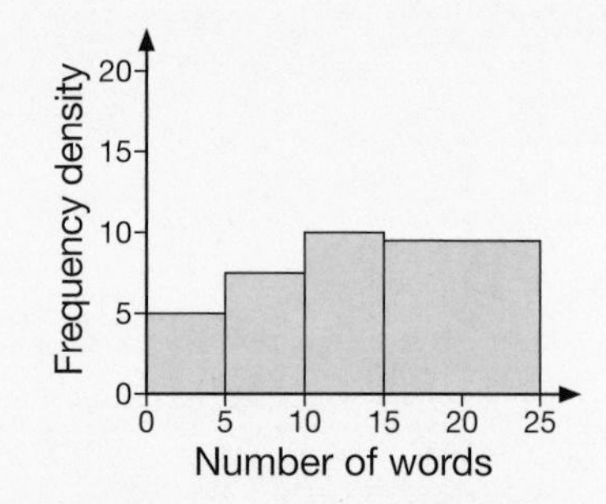

Frequency polygons

A frequency polygon can be drawn from a histogram by joining the midpoints of the tops of the bars with straight lines to form a polygon.

The lines should be extended to the horizontal axis on either side so that the area under the frequency polygon is the same as the area under the histogram.

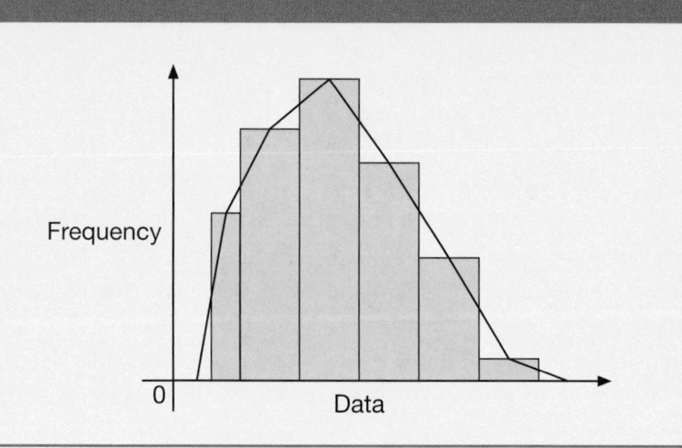

Questions to try

1 One hundred teenagers were asked how much television they had watched the previous evening.

The results of the survey are shown in the table below.

Time t (hours)	Frequency
$0 \leqslant t < 0.5$	16
$0.5 \leqslant t < 1$	28
$1 \leqslant t < 2$	26
$2 \leqslant t < 3$	12
$3 \leqslant t < 5$	18

Draw a histogram to illustrate the results.

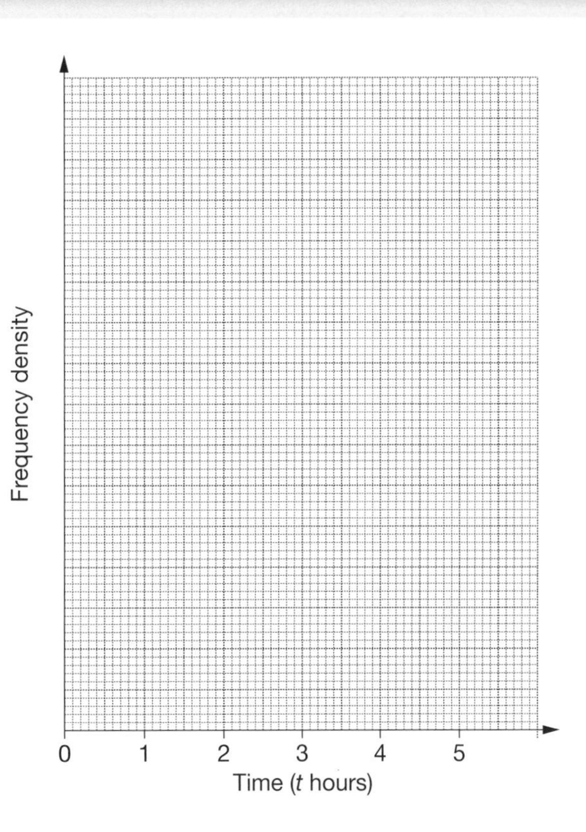

2 The speeds of 100 cars travelling along a road are shown in the table.

Speed (s km/h)	$20 \leqslant s < 35$	$35 \leqslant s < 45$	$45 \leqslant s < 55$	$55 \leqslant s < 65$	$65 \leqslant s < 85$
Frequency	6	19	34	26	15

(a) Draw a histogram to show this information.

(b) The speed limit along this road is 48 km/h. Estimate the number of cars exceeding the speed limit.

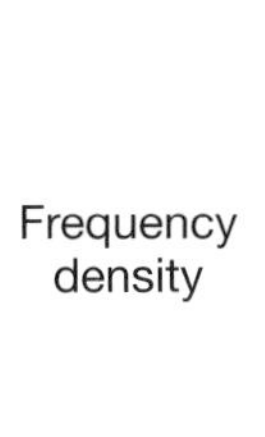

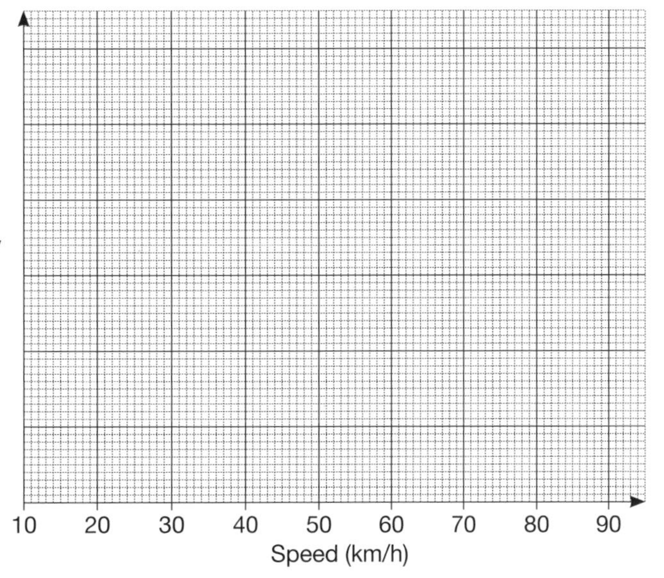

3 The frequency distribution shows the heights of 50 plants measured to the nearest centimetre.

Draw a frequency polygon to represent this information.

Height (cm)	Frequency
15–19	4
20–24	6
25–29	7
30–39	11
40–49	9
50–74	5
75–99	2

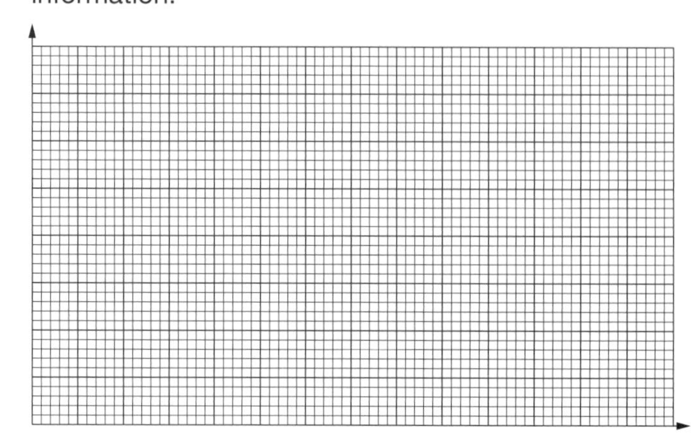

4 The table gives the prices (in pounds) of all 53 models of digital cameras available in British shops in December 1997.

(a) Estimate the mean price of the cameras.

(b) Draw a histogram to illustrate the prices in December 1997.

Price (£p)	Frequency
$0 \leqslant p < 200$	6
$200 \leqslant p < 400$	21
$400 \leqslant p < 600$	17
$600 \leqslant p < 1000$	7
$1000 \leqslant p < 2000$	2

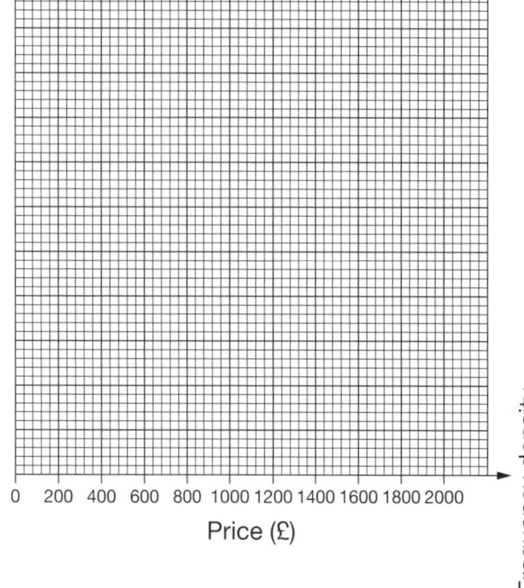

The histogram below shows the prices of all digital cameras in British shops in May 1997.

(c) How many cameras were available in May 1997?

(d) Compare, in two ways, the prices of cameras in December 1997 with the prices in May 1997.

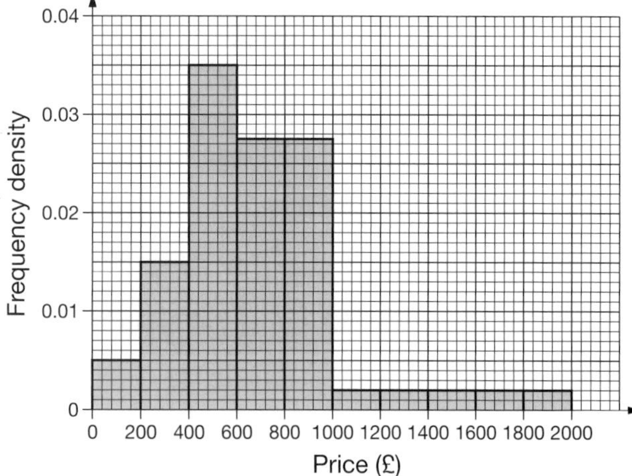

5 A company manufactures various types of lightbulbs. One type of bulb manufactured by the company is considered to be environmentally friendly since bulbs of this type have a long lifetime.

The lifetimes, in thousands of hours, of a sample of 50 bulbs of this type are measured and the results are summarised in the grouped frequency distribution on the right.

Lifetime, x (1000s hours)	Number of bulbs, f	Frequency density
$6 \leqslant x < 10$	4	
$10 \leqslant x < 12$	15	
$12 \leqslant x < 14$	19	
$14 \leqslant x < 16$	8	
$16 \leqslant x < 20$	4	

(a) Complete the frequency density column in the table and draw a histogram of the data in the table.

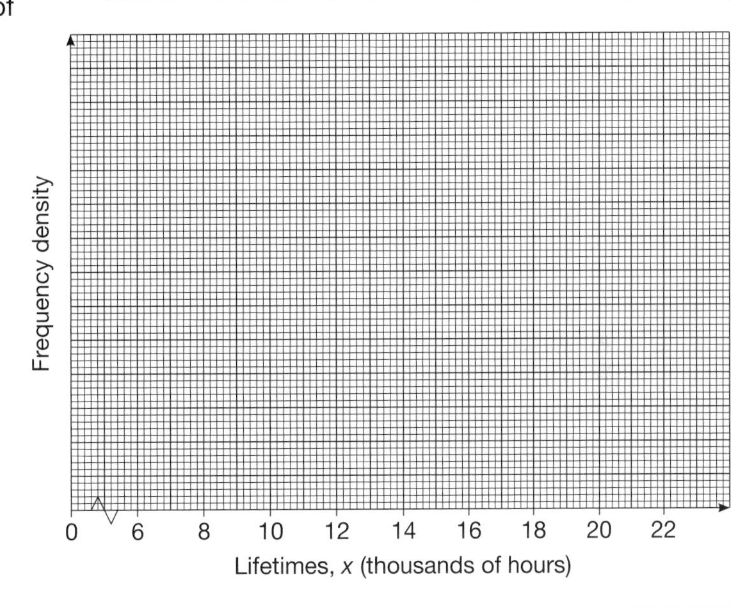

(b) Another company also manufactures an environmentally-friendly light bulb. The lifetimes, in thousands of hours, of a sample of these bulbs are also measured. The results are shown in the histogram on the right.

Estimate the proportion of the light bulbs manufactured by the second company that last longer than 15 000 hours.

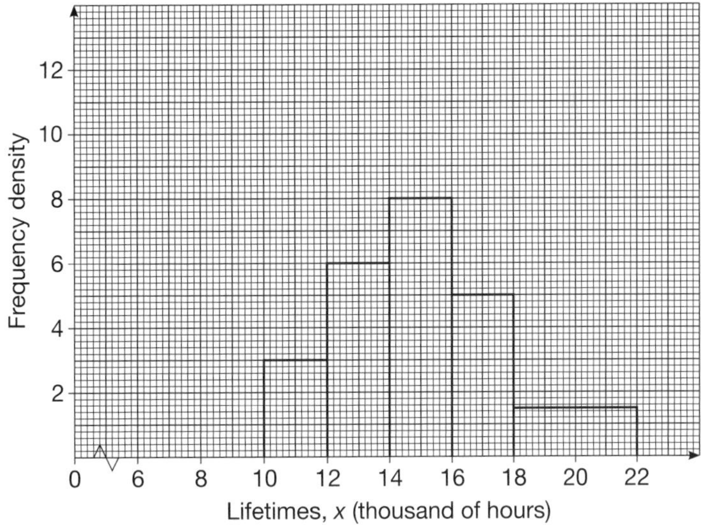

(c) By comparing the two histograms, and giving a reason for your answer, comment on which of the two companies manufactures the most environmentally-friendly light bulbs.

17 Further probability

Exam Question and Answer

1 A bag contains five red and four blue counters. A counter is drawn from the bag and then a second counter is drawn from the bag. Draw a tree diagram to show the various possibilities that can occur and use the diagram to find the probability that both counters are blue.

Answer

With replacement:

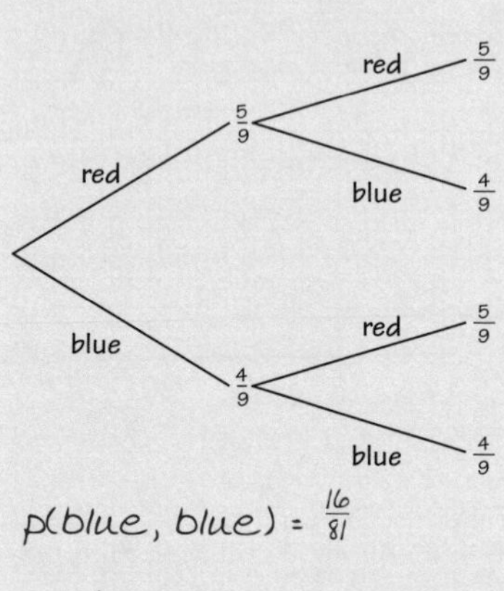

p(blue, blue) = $\frac{16}{81}$

Without replacement:

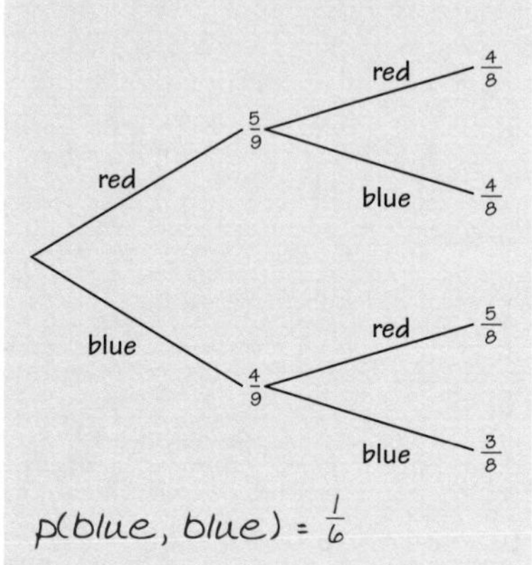

p(blue, blue) = $\frac{1}{6}$

How to score full marks

The question does not make it clear whether the first counter is replaced before the second counter is drawn. This gives rise to two possibilities.

With replacement (independent events)

- Where the first counter is replaced before the second counter is drawn then the two events are independent and the probabilities for each event are the same.
- You should have a diagram as shown on the left.
- From the diagram, the probability that both counters are blue, p(blue, blue), is p(blue counter drawn first and blue counter drawn second)

  = p(blue counter drawn first) × p(blue counter drawn second)

  $= \frac{4}{9} \times \frac{4}{9}$

  $= \frac{16}{81}$

Without replacement (dependent events)

- Where the first counter is not replaced before the second counter is drawn then the two events are not independent (i.e. they are dependent) and the probabilities for the second event will be affected by the outcomes of the first event.
- For the second counter then:

  if the first counter was red, there are still 4 blue counters and 8 counters altogether

  if the first counter was blue, there are now 3 blue counters and 8 counters altogether.
- You should have a diagram as shown on the left.
- From the diagram you can see that the probability that both counters are blue, p(blue, blue)
  = p(blue counter drawn first and blue counter drawn second)
  = p(blue counter drawn first) × p(blue counter drawn second)

  $= \frac{4}{9} \times \frac{3}{8}$

  $= \frac{12}{72}$

  $= \frac{1}{6}$

Key points to remember

Probability

In probability, an event that is certain to happen has a probability of 1 whereas an event that is impossible has a probability of 0.

In general:

$$\text{p(success)} = \frac{\text{number of 'successful' outcomes}}{\text{number of possible outcomes}}$$

where p(success) means the probability of success.

Addition rule – mutually exclusive events

Events are mutually exclusive if they cannot happen at the same time.

For mutually exclusive events, use the addition rule (also called the **or** rule):

P(A or B) = p(A) + p(B)

Remember that to add fractions you must first find a common denominator.

Multiplication rule – independent events

Events are **independent** if one event does not affect the other.

For independent events, use the multiplication rule (also called the **and** rule):

p(A and B) = p(A) × p(B)

Remember that to multiply fractions you multiply the numerators and you multiply the denominators.

DON'T MAKE THESE MISTAKES ...

✗ Don't discount zeros in decimal fractions!

$0.1 \times 0.1 = 0.01$

$0.2 \times 0.2 = 0.04$

$0.3 \times 0.3 = 0.09$ …

✗ **Don't forget**

Probabilities greater than 1 or less than 0 do not have any meaning.

H

Questions to try

1 A bag contains four red and six blue balls. One ball is chosen at random and its colour is noted. The ball is not replaced and a second ball is chosen at random and its colour is noted. Draw a tree diagram to represent this situation and use it to calculate:

(a) the probability of obtaining two red balls

(b) the probability of obtaining one ball of each colour.

2 A box contains five red discs, four blue discs, two white discs and one black disc. Three discs are drawn at random without replacement from the box.

Calculate the probability that:

(a) the three discs drawn are red

(b) the three discs drawn are of the same colour

(c) exactly two of the discs drawn are red.

3 Q, R and S are three mutually exclusive events where p(event Q) = 30%, p(event R) = $\frac{1}{4}$ and p(event S) = 0.2.

Calculate the probability of:

(a) either Q or R occurring

(b) either R or S occurring

(c) any of Q or R or S occurring.

4 A biased six-sided die is such that a score of 6 is twice as likely as a score of 5. A score of 5 is twice as likely as a score of 4. The probabilities of a score of 1, 2, 3 and 4 are all the same.

Calculate the probability of:

(a) a score of 6

(b) a score of 5

(c) a score less than 4.

5 Some students decide to organise an evening out.
They can only go on a Friday or Saturday.

$\frac{5}{6}$ of the students choose bowling.

The rest of the students choose skating.

$\frac{2}{5}$ of those choosing bowling prefer Friday.

$\frac{3}{8}$ of those choosing skating prefer Saturday.

(a) One student is chosen at random.
Calculate the probability that this student prefers Saturday.

(b) 90 students prefer Saturday.
How many students are there altogether?

 The answers can be found on pages **119–120.**

Exam Questions and Answers

1 (a) Write down:

(i) the 8th term and

(ii) the *n*th term of this sequence.

1 5 9 13 17 …

(b) Use your answer to (a) to write down the *n*th term of this sequence.

1 25 81 169 289 …

Answer

(a) (i) 29

(ii) *n*th term $= 4n - 3$

(b) *n*th term $= (4n - 3)^2$

How to score full marks

- You should be able to find the 8th term quite easily, by counting on in the pattern (add 4 each time).

  1 5 9 13 17 21 25 29

- The sequence is formed by adding on the same number each time, so the first differences are all the same. This means the sequence is linear. As each term is found by 'adding on 4', the *n*th term must have something to do with $4n$. The first term is 1, so the rule is $4n - 3$.
- The question asks you to use the first sequence. If you look carefully, you will see that each term of the second sequence is the square of the equivalent term in the first sequence.
- The *n*th term is therefore the square of the *n*th term in the first sequence.

2

2 (a) A sequence of patterns is shown.

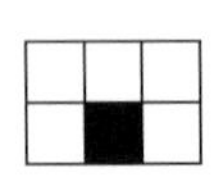
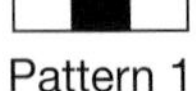

Pattern 1

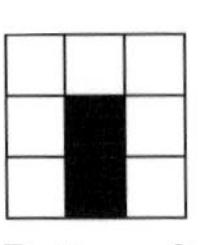

Pattern 2

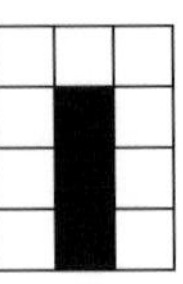

Pattern 3

Write an expression, in terms of n, for the number of white squares in the nth pattern of the sequence.

(b) A number sequence begins:

3, 6, 11, 18, 27, ...

Write an expression, in terms of n, for the nth term of this sequence.

2

Answer

(a) nth term $= n + n + 3$
$= 2n + 3$

(b) nth term $= n^2 + 2$

How to score full marks

► You can use the geometrical properties of the patterns to find the solution.

For example the 4th pattern (pattern 4) gives:

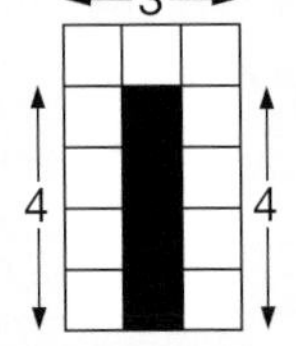

and the 5th pattern (pattern 5) gives:

so the nth pattern gives:

i.e. $n + n + 3$ or $2n + 3$

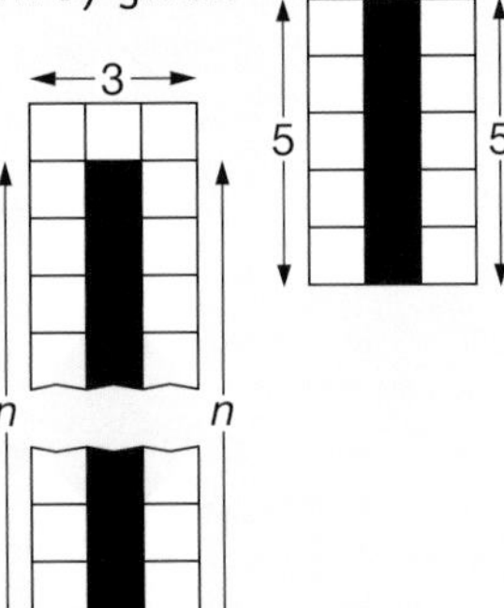

► Look at the sequence.

Sequence	3		6		11		18		27		...
1st difference		+3		+5		+7		+9		...	
2nd difference			+2		+2		+2		...		

► As the first differences are not the same, you need to work out the second differences. These are the same so the sequence is quadratic.

n^2	1	4	9	16	25	
Sequence	3	6	11	18	27	...

► Each term of the sequence is equal to $n^2 + 2$ so this is the nth term.

Key points to remember

Definition: A **sequence** is a set of numbers that follow a particular rule.

Linear sequences

For the sequence 3, 7, 11, 15, 19, … you can work out the differences as shown.

Sequence	3	7	11	15	19	…
First difference	+4	+4	+4	+4	…	

The first differences are all the same, so the sequence is linear and you can use:

nth term = first term + $(n - 1) \times$ first difference
$= 3 + (n - 1) \times 4$
$= 3 + 4n - 4$
$= 4n - 1$

Quadratic sequences

Quadratic sequences will be limited to variations on n^2.

Sequence	nth term
1, 4, 9, 16, …	n^2
2, 5, 10, 17, …	$n^2 + 1$
0, 3, 8, 15, …	$n^2 - 1$
2, 8, 18, 32, …	$2n^2$

DON'T MAKE THESE MISTAKES …

- ✗ Don't forget that the **nth term must work** for **all of the given values**, not just the first one or two.
- ✗ **Don't jump in and use the formulae** for linear and quadratic equations before you have tried to find an obvious simple answer to the sequence.
- ✗ **Don't miss patterns** where some numbers seem to be missing from the sequence, e.g. 4, 9, 16, 25, …
  This is based on the square numbers but the nth term is not n^2 but $(n + 1)^2$.
- ✗ **Don't write out individual values** for sequences in diagram form – they can usually be solved by looking at the geometrical shapes in the diagram.

Questions to try

1 Write down the first five terms of a sequence where the *nth* term is given as:

(a) $2n - 1$ (b) $n^2 - n$ (c) $\frac{n}{n+1}$

2 (a) What is the next number in this sequence?

3, 7, 11, 15, … .

One number in the sequence is x.

(b) (i) Write, in terms of x, the next number in the sequence.

(ii) Write, in terms of x, the number in the sequence before x.

3 (a) Write down the next two terms of the following sequence.

2, 5, 11, 23, 47, …

(b) Explain how you get from one term of the sequence to the next term.

(c) Write down an expression for the nth term of this sequence.

$\frac{1}{4}, \frac{2}{5}, \frac{3}{6}, \frac{4}{7}, \frac{5}{8}, \ldots$

4 (a) Write down the first three terms of the sequence whose nth term is given by

$$\frac{5n}{2n+9}$$

(b) Which term of the sequence has a value of 2?

5 (a) A sequence begins 2, 5, 8, 11, … .

Write an expression for the nth term of the sequence.

(b) The nth term of a different sequence is $\frac{1}{2}(n^2 - 2n)$.

(i) Write down the first term and the second term of the sequence.

(ii) When the nth term of the sequence is 40, the value of n can be found by solving the equation $n^2 - 2n - 80 = 0$.

Solve this equation and hence find the value of n.

You must show all your working.

6 Study the patterns of black and white squares shown below.

(a) How many white squares are there in the 4th pattern?

(b) Write an expression, in terms of n, for the number of white squares in the nth pattern.

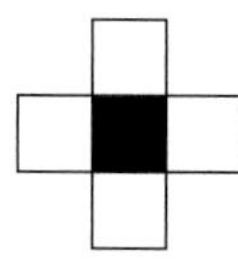

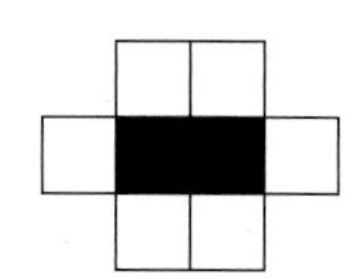

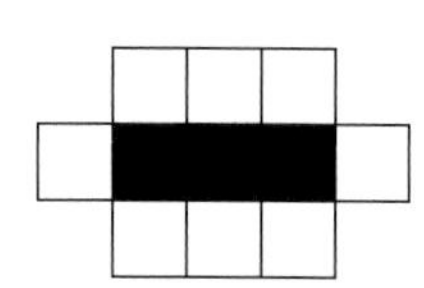

7 A sequence of matchstick patterns is shown below.

Write an expression, in terms of n, for the number of matchsticks in the nth pattern of the sequence.

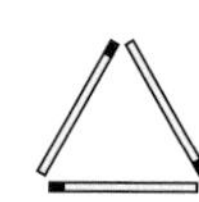

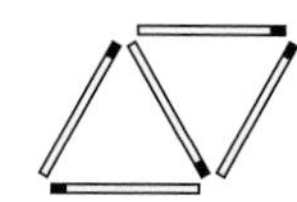

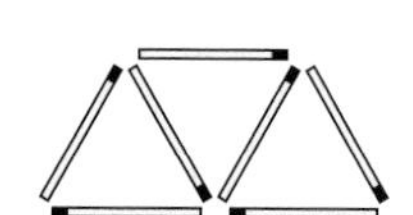

The answers can be found on pages **120–121.**

Manipulation of equations 19

Exam Questions and Answers

1 Make l the subject of the formula $T = 2\pi\sqrt{\frac{l}{g}}$.

Answer

$$T = 2\pi\sqrt{\frac{l}{g}}$$

$$\frac{T}{2\pi} = \sqrt{\frac{l}{g}}$$

$$\left(\frac{T}{2\pi}\right)^2 = \frac{l}{g}$$

$$g\left(\frac{T}{2\pi}\right)^2 = l$$

$$l = g\left(\frac{T}{2\pi}\right)^2$$

How to score full marks

- Divide both sides of the formula by 2π.
- Square both sides (to get rid of the √ sign).
- Multiply both sides of the formula by g to isolate l.
- Turn the formula around so that l is the subject.

2 Solve the following equation.

$$\frac{2x-3}{6} + \frac{x+2}{3} = \frac{5}{2}$$

Answer

$$\frac{2x-3}{6} + \frac{x+2}{3} = \frac{5}{2}$$

$$\frac{2x-3}{6} + \frac{2(x+2)}{6} = \frac{15}{6}$$

$$2x - 3 + 2(x+2) = 15$$

$$2x - 3 + 2x + 4 = 15$$

$$4x + 1 = 15$$

$$4x = 14$$

$$x = 3.5 \text{ or } 3\tfrac{1}{2}$$

How to score full marks

- To solve questions involving algebraic fractions, you need to change all the terms to fractions with the same (common) denominator.
- $\frac{2x-3}{6} + \frac{x+2}{3} = \frac{5}{2}$ $\quad \frac{x+2}{3} = \frac{2(x+2)}{6}$ $\quad \frac{5}{2} = \frac{15}{6}$

  $\frac{2x-3}{6} + \frac{2(x+2)}{6} = \frac{15}{6}$ (writing all the terms with a denominator of 6)

  $2x - 3 + 2(x+2) = 15$ (multiplying throughout by 6)

  $2x - 3 + 2x + 4 = 15$ (simplifying)

  $4x + 1 = 15$ (simplifying further)

  $4x = 14$

  $x = \frac{14}{4} = 3.5$ or $3\frac{1}{2}$ (solving to find the value of x)

Key points to remember

Doing and undoing

Every mathematical operation has an inverse. You can use the inverse to 'undo' an equation or formula.

Operation	Inverse		
add 5	subtract 5	$a + 5 = 7$ $a + 5 - 5 = 7 - 5$ $a = 2$	(as $+5 - 5 = 0$ and $a + 0 = a$)
subtract 3	add 3	$b - 3 = 9$ $b - 3 + 3 = 9 + 3$ $b = 12$	(as $-3 + 3 = 0$ and $b + 0 = b$)
multiply by 7	divide by 7 (or multiply by $\frac{1}{7}$)	$c \times 7 = 21$ $c \times 7 \div 7 = 21 \div 7$ $c = 3$	(as $7 \div 7 = 1$ and $c \times 1 = c$)
divide by 4	multiply by 4	$d \div 4 = 11$ $d \div 4 \times 4 = 11 \times 4$ $d = 44$	(as $\div 4 \times 4 = 1$ and $d \times 1 = d$)
square	take the square root	$e^2 = 49$ $\sqrt{e^2} = \sqrt{49}$ $e = 7$	
take the square root	square	$\sqrt{f} = 10$ $(\sqrt{f})^2 = 10^2$ $f = 100$	

Rearranging formulae

You can rearrange (or transpose) a formula in exactly the same way as you solve an equation. You must always keep the formula balanced, so you must make sure that whatever you do to one side, you also do to the other side.

For the formula $S = \frac{D}{T}$, S is the subject of the formula.

The formula can be rearranged to make D or T the subject.

$S = \frac{D}{T}$

$D = S \times T$ or

$D = ST$ (multiplying both sides of the formula by T)

Or, from $D = ST$:

$\frac{D}{S} = T$ (dividing both sides of the formula by S)

$T = \frac{D}{S}$ (turning the formula around)

Algebraic fractions

If you remember how to work with numerical fractions, you will be able to use the same methods to work out algebraic fractions.

For example:

$\frac{1}{3} + \frac{1}{5} + \frac{1}{6}$

To add these fractions, first express them all as equivalent fractions with a common denominator.

Remember that to make an equivalent fraction you multiply or divide the numerator and the denominator by the same number.

So $\frac{1}{3} + \frac{1}{5} + \frac{1}{6} = \frac{10}{30} + \frac{6}{30} + \frac{5}{30} = \frac{21}{30}$

DON'T MAKE THESE MISTAKES ...

✗ To add fractions you must **always make sure they are over a common denominator.**

✗ **You can always find a common denominator by multiplying all the denominators together**, but this denominator will not always be the lowest common denominator.

$\frac{1}{3} = \frac{10}{30}$ $\frac{1}{5} = \frac{6}{30}$ $\frac{1}{6} = \frac{5}{30}$

Questions to try

1 Rearrange the following formula to make p the subject:

$$t = \frac{p + q}{2}$$

2 Rearrange the formula $r = 4pq - s$ to make:

(a) s the subject

(b) q the subject.

3 A small paving slab weighs x kilograms.

A large paving slab weighs $(2x + 3)$ kilograms.

(a) Write an expression, in terms of x, for the total weight of 16 small slabs and 4 large slabs.

Give your answer in its simplest form.

The total weight of the slabs is 132 kilograms.

(b) Write down an equation and find the value of x.

4 The diagram shows a rectangle with length $3x + 2$ and width $2x$.

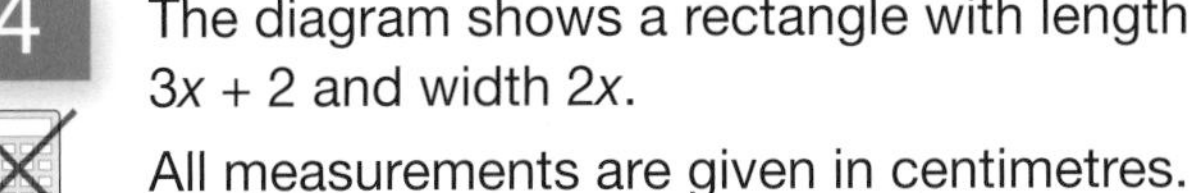

All measurements are given in centimetres.

The perimeter of the rectangle is P centimetres.

The area of the rectangle is A square centimetres.

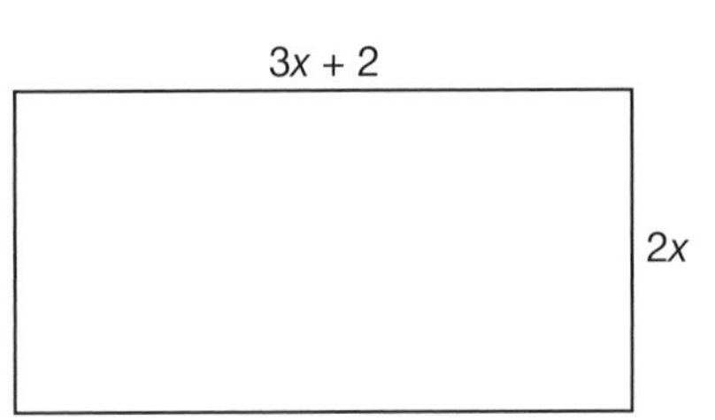

Not drawn accurately

(a) Write down an expression in its simplest form, in terms of x, for:

(i) P (ii) A.

The perimeter of the rectangle is 44 cm.

(b) Work out the value of A.

5 The triangle has angles $x°$, $2x°$ and $81°$ as shown.

Find the value of x.

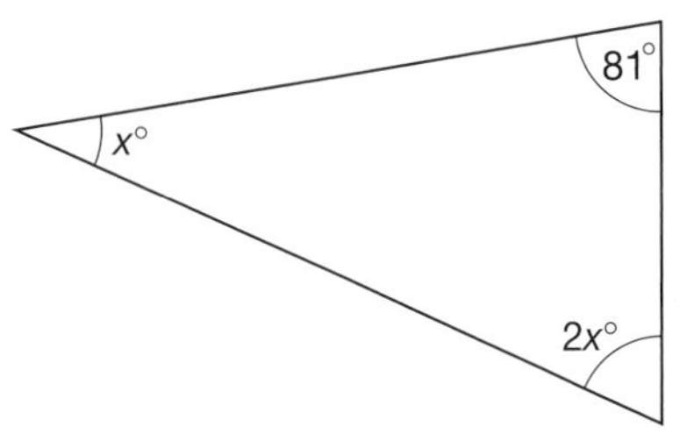

Not drawn accurately

6 Neil buys 30 shares in a company. Each share costs £x.

(a) How much does Neil spend?

(b) He sells 10 shares when the price of each share has risen by £1.

How much does he receive?

(c) He sells the other 20 shares when the price of each share is double what he paid for it. How much does he receive for these 20 shares?

(d) Neil makes £70 profit.

Form an equation in x and solve it to find the original cost of a share.

The answers can be found on page **122**.

20 Simultaneous equations

Exam Questions and Answers

1 **1** Solve these simultaneous equations.

$$x + 2y = 7$$
$$3x - 2y = 5$$

1 **Answer**

$$x + 2y = 7$$
$$3x - 2y = 5$$

$$4x = 12$$
$$x = 3$$

$$x + 2y = 7$$
$$3 + 2y = 7$$
$$2y = 4$$
$$y = 2$$

How to score full marks

- Where the terms in x or the terms in y are numerically equal, you need to use the method of elimination (eliminate the term in x or y by adding or subtracting the two equations).
- In this example the terms in y are numerically equal (and the signs are opposite) so add the two equations.
- You need to divide both sides by 4 to find the value of x.
- Substitute the value of x (= 3) into the first equation.
- Subtract 3 from both sides.
- Divide both sides by 2 to find the value of y.

2 **2** Solve these simultaneous equations.

$$x + 4y = 11$$
$$2x + y = 1$$

2 **Answer**

$$x + 4y = 11$$
$$2x + y = 1$$

$$x + 4y = 11$$
$$x = 11 - 4y$$

$$2x + y = 1$$
$$2(11 - 4y) + y = 1$$
$$22 - 8y + y = 1$$
$$22 - 7y = 1$$

$$22 = 1 + 7y$$
$$21 = 7y$$
$$y = 3$$

$$x + 4y = 11$$
$$x + 4 \times 3 = 11$$
$$x + 12 = 11$$
$$x = -1$$

How to score full marks

- Where neither the terms in x nor the terms in y are numerically equal, you need to use the method of substitution (by making x or y the subject of one of the equations).
- Use the first equation to make x the subject, by subtracting $4y$ from both sides.
- Replace the value of x in the second equation by $11 - 4y$.
- Expand the left-hand side of the equation, then simplify it by collecting like terms to find the value of y.
- Substitute the value of y (= 3) into the first equation.
- Simplify it and solve it to find the value of x.

Key points to remember

You can solve simultaneous equations by **elimination or substitution** – think about which method you are going to use, before you start.

- Where the terms in x or the terms in y are numerically equal, use elimination to remove the term in x or y by adding or subtracting the two equations.
- Where neither the terms in x nor the terms in y are numerically equal, use the method of substitution by making x or y the subject of one of the equations.

You need to know these methods as you will not always be able to use a calculator, and trial and error methods rarely receive full marks (especially if you only give one solution).

For the **elimination method**, you need to add the equations if the signs are opposite and subtract the equations if the signs are the same.

For the **substitution method**, choose the subject carefully (preferably where there is a single term in x or a single term in y) even if this means using the second equation.

If you get **silly answers**, such as very large numbers or difficult fractions, **go back and check your working.**

Check your answers by **substituting them back into both equations** to make sure that you have not made any silly mistakes.

DON'T MAKE THESE MISTAKES ...

✗ **Don't panic if the examination paper uses different letters.**

$a + 2b = 7$

$3a - 2b = 5$

Use the method described above but remember the answers will be $a = \ldots$ and $b = \ldots$

✗ **When using the elimination method, watch out for negative signs.**

$4x - y = 7$

$4x - 2y = 4$

gives:

$-y - (-2y) = 7 - 4$

Use brackets to help you.

$-y + 2y = 3$

$y = 3$

Questions to try

1 Solve these simultaneous equations.

$2x + 3y = 7$

$5x - 3y = 21$

2 Solve these simultaneous equations.

$5x + 3y = 19$

$5x - 2y = 4$

3 Solve these simultaneous equations.

$2x + 3y = 1$

$x - 2y = 11$

4 Two numbers have a sum of 13 and a difference of 3.
What are the two numbers?

5 Calculate the coordinates of the point of intersection of these lines.

$x + y = 8$

$y = 3x - 4$

The answers can be found on pages **123–124.**

Exam Question and Answer

1 The diagram shows a square ABCD and a rectangle RBPQ.

CP = 2 cm and RA = 3 cm.

The length of the side of the square ABCD is x cm.

The rectangle RBPQ has an area of 42 cm^2.

Form an equation, in terms of x, for the area of the rectangle. Show that it can be written in the form $x^2 + 5x - 36 = 0$.

Solve $x^2 + 5x - 36 = 0$, and hence calculate the area of the square ABCD.

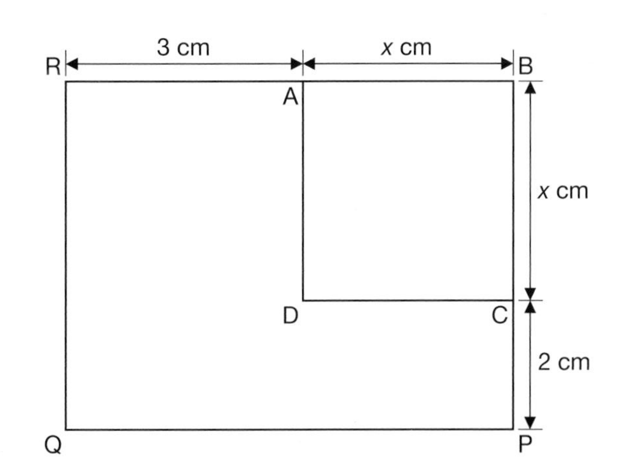

Answer

Area of RBPQ = $(3 + x)(x + 2)$

$(3 + x)(x + 2) = 42$

$3x + 6 + x^2 + 2x = 42$

$x^2 + 5x + 6 = 42$

$x^2 + 5x + 6 - 42 = 0$

$x^2 + 5x - 36 = 0$

$x^2 + 5x - 36 = 0$

$(x + 9)(x - 4) = 0$

$x + 9 = 0$ so $x = -9$

or $x - 4 = 0$ so $x = 4$

The side of the square is 4 cm (since you cannot have a length of −9 cm).

How to score full marks

► You are given that the rectangle RBPQ has an area of 42 cm^2 so find an algebraic equation for this, and solve it.

Area of rectangle = $(3 + x)(x + 2) = 42$

Start by expanding the brackets (remember FOIL), then simplify to get the form $x^2 + 5x - 36 = 0$ which is given in the question.

► Now solve this quadratic to produce two solutions of $x = -9$ and $x = 4$. As you cannot have a length of −9 cm, the length of the square is 4 cm.

► It is always a good idea to explain why you have rejected one of your solutions so that the examiner appreciates how clever you are.

Key points to remember

Expanding quadratic expressions

You can use FOIL to remind you how to find the product of two binomial expressions.

F = First
O = Outsides
I = Insides
L = Last.

Product

F = First	$(x+3)(2x-5)$	$x \times 2x$	$= 2x^2$
O = Outsides	$(x+3)(2x-5)$	$x \times -5$	$= -5x$
I = Insides	$(x+3)(2x-5)$	$+3 \times 2x$	$= 6x$
L = Last	$(x+3)(2x-5)$	$+3 \times -5$	$= -15$

Factorising quadratic equations

Remember that if the product of two numbers is zero, then one or both of them must be zero.

If $ab = 0$ then either $a = 0$ or $b = 0$ or both $a = 0$ and $b = 0$.

You can use this fact to solve quadratic equations.

If $(2x - 5)(x + 3) = 0$ then either $(2x - 5) = 0$ or $(x + 3) = 0$ or both $(2x - 5) = 0$ and $(x + 3) = 0$.

If $(2x - 5) = 0$ then $x = \frac{5}{2}$ and if $(x + 3) = 0$ then $x = -3$.

Factorising quadratic expressions

You can use the reverse process to write a quadratic expression as the product of two binomial expressions.

$$\begin{aligned} 2x^2 + x - 15 &= 2x^2 - 5x + 6x - 15 \\ &= x \times 2x - 5 \times x + 3 \times 2x - 3 \times 5 \\ &= x \times (2x - 5) + 3 \times (2x - 5) \\ &= (2x - 5)(x + 3) \end{aligned}$$

The difficult part of this method is working out how to rewrite the three terms as four terms, to find two brackets.

You need to work logically, using your common sense, to factorise quadratic expressions.
You know that $2x^2 + x - 15 = (2x \quad)(x \quad)$ since $2x \times x = 2x^2$ as required.

Now you need to look for pairs of numbers that multiply together to give -15.

DON'T MAKE THESE MISTAKES ...

✗ **Don't forget to give both solutions**. Quadratic equations should always give you two solutions **but**:

- some quadratic equations cannot be solved e.g. $x^2 + x + 1 = 0$
- some quadratic equations give a repeated root e.g. $x^2 - 4x + 4 = 0$

✗ **You can factorise** $x^2 - 4x + 4 = 0$ as $(x - 2)(x - 2) = 0$ **to find the solutions** $(x - 2) = 0$ or $(x - 2) = 0$, i.e. $x = 2$ (twice).

✗ When you are asked to **solve a quadratic equation** you must always give both solutions.

✗ **Don't try to find the solutions by guesswork (trial and improvement)**.

Questions to try

1 Expand and simplify these expressions.

(a) $(x + 2)(x + 7)$ (b) $(3x - 2)(4x - 7)$

(c) $(2x - 1)^2$

2 Factorise these expressions completely.

(a) $x^2 + x - 12$ (b) $x^2 - x - 12$

(c) $x^2 + x - 3$

3 Solve the following quadratic equations.

(a) $(x - 4)(2x - 7) = 0$ (b) $x^2 - 7x + 12 = 0$

(c) $2x^2 - 5x - 3 = 0$ (d) $2x^2 - 5x = 3$

4 (a) Factorise $x^2 - 9x + 20$.

(b) Solve $x^2 + 2x = 0$.

5 Given that $(3x - 1)(x + 2) = 3x^2 + ax + b$, find a and b.

6 (a) Expand and simplify $(x + y)^2$.

(b) Hence or otherwise find the value of $3.63^2 + 2 \times 3.63 \times 1.37 + 1.37^2$.

7 Expand $(a + b)(a - b)$ and use this to work out $1999^2 - 1998^2$.

8 Simplify the expression $(x + 8)^2 - (x - 8)^2$.

9 The diagram shows a square ABCD with a triangle DCE, right-angled at D, placed so that the side DC is common.

The length of the side of the square is x cm and DE = 4 cm. The area of the whole figure ABCE is 48 cm^2.

(a) Show clearly that x satisfies the equation $x^2 + 2x = 48$.

(b) Solve the equation and write down the length of the side of the square.

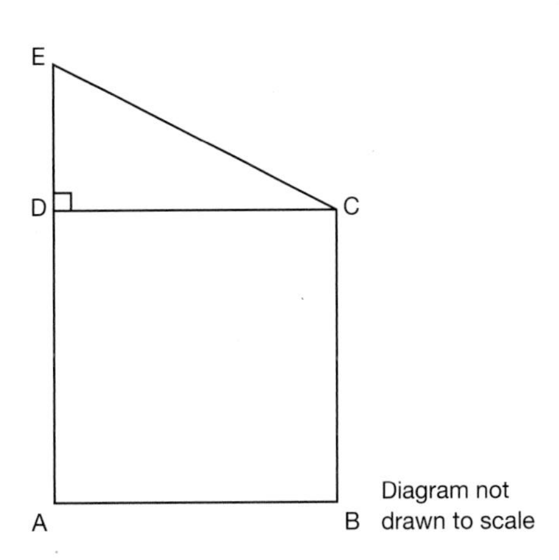

Diagram not drawn to scale

The answers can be found on pages **124–125**.

22 Further manipulation of equations

Exam Questions and Answers H

1 **1** Make g the subject of the following formula.

$$\frac{t(3+g)}{8-g} = 2$$

1 Answer

$$\frac{t(3+g)}{8-g} = 2$$

$t(3 + g) = 2(8 - g)$

$3t + tg = 16 - 2g$

$tg = 16 - 2g - 3t$

$tg + 2g = 16 - 3t$

$g(t + 2) = 16 - 3t$

$$g = \frac{16-3t}{t+2}$$

How to score full marks

- Use these steps to change the subject of the formula.
- First remove the fractions by multiplying both sides by $(8 - g)$.
- Expand the brackets.
- Subtract $3t$ from the left-hand side to get all terms in g on one side.
- Factorise, then divide both sides by $(t + 2)$ to make g the subject.

2 **2** A cyclist travels 30 miles at an average speed of x miles per hour, followed by 20 miles at $(x + 5)$ miles per hour.

(a) Write down an expression, in terms of x, for the time taken, in hours, for the whole journey.

The time taken for the whole journey is 3 hours.

(b) Write down an equation, in terms of x, for the time taken, in hours, for the whole journey and show that this simplifies to $3x^2 - 35x - 150 = 0$.

(c) Hence solve your equation to find the value of x.

Answer

(a) For first 30 miles:

$T = \frac{30}{x}$

For next 20 miles:

$T = \frac{20}{x+5}$

Total time $= \frac{30}{x} + \frac{20}{x+5}$

(b) $\frac{30}{x} + \frac{20}{x+5} = 3$

$30(x+5) + 20x = 3x(x+5)$

$30x + 150 + 20x = 3x^2 + 15x$

$3x^2 - 35x - 150 = 0$

(c) $3x^2 - 35x - 150 = 0$

$(3x+10)(x-15) = 0$

so $3x + 10 = 0$ or $x - 15 = 0$

$x = -\frac{10}{3}$ $\quad$ $x = 15$

The value of x is 15 (as x cannot be negative).

How to score full marks

- First, remember that speed $= \frac{\text{distance}}{\text{time}}$ and time $= \frac{\text{distance}}{\text{speed}}$.

  It is a good idea to write this down so the examiner knows what you are trying to do.
- Next, you need to break down the task by thinking about both parts of the journey separately.
  The total time for the whole journey is the sum of these two times.
  Remember that this is the solution to part (a) of the question.
- Given that the total time is 3 hours then $\frac{30}{x} + \frac{20}{x+5} = 3$.
- Remember that to solve questions involving algebraic fractions, you need to change all of the terms to fractions with the common denominator.
- You can put all of the terms over a common denominator of $x(x + 5)$:

$$\frac{30}{x} = \frac{30(x+5)}{x(x+5)}$$

$$\frac{20}{x+5} = \frac{20x}{x(x+5)}$$

$$3 = \frac{3x(x+5)}{x(x+5)}$$

- Now multiply throughout by $x(x + 5)$:

  $30(x + 5) + 20x = 3x(x + 5)$
- Expand the brackets and simplify to form the required quadratic equation.

  $3x^2 - 35x - 150 = 0$
- Even if you can't prove the required quadratic equation, you can still gain marks for part (c) using the fact that the quadratic should simplify to $3x^2 - 35x - 150 = 0$.

  You can solve it to find the value of x.
- You should only give one value of x as the other is not suitable. It is always a good idea to check whether all of your solutions are valid.

H

Key points to remember

Rearranging formulae

At the Higher level you may be asked to rearrange formulae where the subject appears in more than one term. Most of these types of question can be solved using the following flow chart.

Take out fractions by finding the common denominator and multiplying both sides by it. → Expand the brackets and take all terms that include the subject to one side. → Take the subject out, as a common factor, and divide both sides by the factor that multiplies it.

The following example will help to illustrate this.

Make x the subject of the formula $p = \frac{xy}{x - y}$.

Take out the fractions by multiplying both sides by the denominator $(x - y)$.

$$p(x - y) = xy$$

Expand the brackets and take the subject to one side.

$$px - py = xy$$
$$px = xy + py$$
$$px - xy = py$$

Factorise terms containing the subject and divide both sides by the factor that multiplies it.

$$x(p - y) = py$$
$$x = \frac{py}{p - y}$$

Algebraic fractions

Remember how to work with numerical fractions. You can use the same methods to solve algebraic fractions.

To add or subtract fractions, change them all to equivalent fractions with a common denominator.

Common denominator

A common denominator is a number that can be divided exactly by all the denominators of the fractions you are working with.

DON'T MAKE THESE MISTAKES ...

✗ Don't try to **add or subtract algebraic fractions** unless they have common denominators.

✗ **Don't despair!** You can always **find a common denominator** by multiplying the denominators together. So $\frac{1}{x+1}$, $\frac{1}{x+2}$ and $\frac{1}{2x+1}$ will have a common denominator of $(x + 1)(x + 2)(2x + 1)$. Phew!!

Questions to try

1 Make y the subject of the formula $q = \frac{xy}{x - y}$.

2 A formula states $s = \frac{1 - t^2}{1 + t^2}$.

(a) Find the value of s when $t = \frac{1}{3}$.

(b) Rearrange the formula to make t the subject.

3 Rearrange the formula $p = \sqrt{\frac{s}{s - t}}$ to make s the subject.

4 Simplify fully the expression $\frac{3x^2 + 15x}{2x^2 + 9x - 5}$.

5 You are given that

$(x + a)^2 + b = x^2 - 6x + 13$.

Calculate the values of a and b.

6 Write as a single fraction $\frac{1}{x - 2} + \frac{1}{x + 3}$.

7 Solve the following equation.

$$\frac{3}{x + 2} - \frac{2}{2x - 3} = \frac{1}{7}$$

he answers can be found on pages **126–127**.

23 Further quadratic equations

Exam Questions and Answers H

1 **1** Solve the equation $3x^2 + 2x - 7 = 0$, giving your answers to 2 decimal places.

1 **Answer**

$a = 3, b = 2, c = -7.$

$$x = \frac{-2 \pm \sqrt{2^2 - 4 \times 3 \times -7}}{2 \times 3}$$

$$= \frac{-2 \pm \sqrt{88}}{6}$$

$$x = \frac{-2 + 9.380\,831\,52}{6}$$

$$\text{or } x = \frac{-2 - 9.380\,831\,52}{6}$$

$x = 1.230\,138\,587$ or $-1.896\,805\,253$

$x = 1.23$ (2 d.p.) or $x = -1.90$ (2 d.p.)

How to score full marks

- The question says 'give your answer correct to 2 decimal places'. This is usually a hint to use the formula.
- Comparing $3x^2 + 2x - 7 = 0$, with the general form $ax^2 + bx + c = 0$:

  $a = 3,\ b = 2,\ c = -7.$
- Substitute these values in the formula $x = \dfrac{-b \pm \sqrt{b^2 - 4ac}}{2a}$.
- Be careful when you round $-1.896\,805\,253$ to two decimal places.

2 **2** (a) (i) Factorise $4x^2 - 37x + 9$.

(ii) Hence, or otherwise, solve the equation $4x^2 - 37x + 9 = 0$.

(b) By considering your answers to part (a), find all the solutions of the equation $4y^4 - 37y^2 + 9 = 0$.

2 **Answer**

(a) (i) $4x^2 - 37x + 9$
$= (4x - 1)(x - 9)$

(ii) $(4x - 1)(x - 9) = 0$
$(4x - 1) = 0$ so $x = \frac{1}{4}$
or $(x - 9) = 0$ so $x = 9$

(b) As $x = y^2$, solutions will be $y^2 = \frac{1}{4}$ so $y = \pm\frac{1}{2}$
and $y^2 = 9$ so $y = \pm 3$

How to score full marks

- Use what you know:

  $4x^2 - 37x + 9 = (2x\quad)(2x\quad)$ (as $2x \times 2x = 4x^2$)
  or $4x^2 - 37x + 9 = (4x\quad)(x\quad)$ (as $4x \times x = 4x^2$)
- If $4x^2 - 37x + 9 = 0$ then $(4x - 1)(x - 9) = 0$

  so $(4x - 1) = 0$ or $(x - 9) = 0$

  $x = \frac{1}{4}$ or $x = 9$
- Compare $4x^2 - 37x + 9 = 0$

  with $4y^4 - 37y^2 + 9 = 0$
- You should notice that if you substitute $x = y^2$ into the second equation, you find the equation you have already solved.

  Then the solutions will be:

  $y^2 = \frac{1}{4}$ and $y^2 = 9$
- Take square roots, and remember that there are two solutions in each case.

  $y = \pm\frac{1}{2}$ and $y = \pm 3$

Key points to remember

Definition: A quadratic equation has the form $ax^2 + bx + c = 0$ where a, b and c are rational numbers. $2x^2 + x - 15 = 0$ is a quadratic equation with $a = 2$, $b = 1$ and $c = -15$.

Factorising quadratic equations (1)

Remember that if the product of the two numbers is zero then one or both of them must be zero.

If $ab = 0$ then either $a = 0$ or $b = 0$ or both $a = 0$ and $b = 0$.

You can use this important fact to solve quadratic equations.

If $(2x - 5)(x + 3) = 0$ then either $(2x - 5) = 0$ or $(x + 3) = 0$ or both $(2x - 5) = 0$ and $(x + 3) = 0$.

Note: If $(2x - 5) = 0$ then $x = \frac{5}{2}$ and if $(x + 3) = 0$ then $x = -3$.

So the solutions of $(2x - 5)(x + 3) = 0$ are $x = \frac{5}{2}$ and $x = -3$.

Quadratic equations – solution by formula

If you can't see how to factorise a quadratic, you can use the formula.

For any quadratic of the form $ax^2 + bx + c = 0$: $x = \dfrac{-b \pm \sqrt{b^2 - 4ac}}{2a}$

Note: The formula for solving quadratic equations will be given in the exam, so you don't need to remember it – but you do need to know how to use it correctly.

Factorising quadratic equations (2)

At the Higher tier, not all quadratic equations can be solved using the method described above. You will need to be able to solve quadratics by using the formula and by iteration techniques. In the exam, the question should make it clear which method you should use. If it doesn't, you can use any method you like.

DON'T MAKE THESE MISTAKES ...

✗ Don't forget that quadratic equations should always give you two solutions **but**:

- some quadratic equations cannot be solved e.g. $x^2 + x + 1 = 0$
- some quadratic equations give a repeated root e.g. $x^2 - 4x + 4 = 0$
  $x^2 - 4x + 4 = 0$
  can be factorised as $(x - 2)(x - 2) = 0$
  with solutions $(x - 2) = 0$ or $(x - 2) = 0$
  i.e. $x = 2$ (twice)
- whenever you are asked to solve a quadratic equation you must give both solutions.

✗ Don't forget that the formula

$$x = \frac{-b + \sqrt{b^2 - 4ac}}{2a}$$

is a short way of writing two different solutions:

$$x = \frac{-b + \sqrt{b^2 - 4ac}}{2a} \quad \text{and} \quad x = \frac{-b - \sqrt{b^2 - 4ac}}{2a}$$

Remember that the **whole** of the numerator is divided by $2a$.

The $(b^2 - 4ac)$ of the formula is very important. It is called the **discriminant**.

If $(b^2 - 4ac) = 0$ then the equation has only one (repeated) root.

If $(b^2 - 4ac) < 0$ then the equation has no roots.

If $(b^2 - 4ac)$ is a square number, the quadratic can be solved by factorising (which is almost always easier than using the formula).

✗ **Never, ever** try to find the solutions of a quadratic equation by guesswork (trial and improvement) unless you are told to do so.

H

Questions to try

1 

Solve the equation $2x^2 + 7x = 4$.

2

Solve the equation $3x^2 - 2x = 0$.

3

Solve the equation $\dfrac{2}{x+2} = \dfrac{x-1}{x}$.

4

Write $x^2 + 4x - 9$ in the form $(x + a)^2 - b$.
Hence solve the equation $x^2 + 4x - 9 = 0$, giving your answer in the form $x = a \pm \sqrt{b}$.

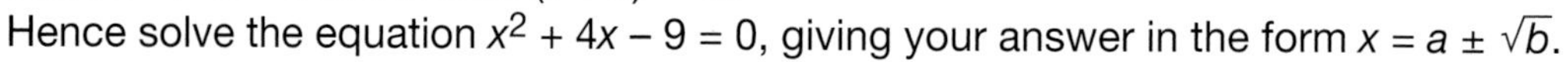

5

Solve $x^2 - 2x - 1 = 0$, giving your answers correct to 2 d.p.

6

(a) Factorise $3x^2 - 2x - 8$.

(b) Solve the equation $3x^2 - 2x - 8 = 0$.

(c) Hence write down all the integers which satisfy $3x^2 - 2x - 8 \leqslant 0$.

The answers can be found on pages **127–128**.

Answers to Questions to try 24

Chapter 1 Transformations

Answer / How to solve these questions

1 **(a)** See the diagram opposite.

(a)

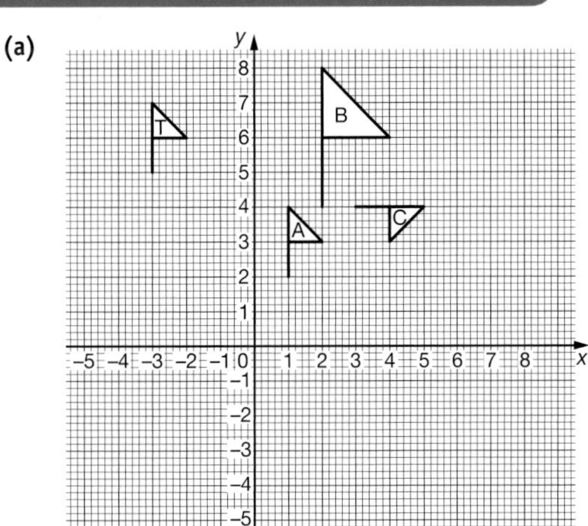

(b) **(i)** The transformation of A onto B is an enlargement, scale factor 2, centre of enlargement (0, 0).

(ii) The transformation of A onto C is a rotation, 90°clockwise, centre of rotation (3, 2).

(b) **(i)** The transformation is an enlargement, so you must give the scale factor (2) and the centre of enlargement (0, 0).

(ii) The transformation is a rotation, so you must give the angle (90°), the direction (clockwise) and the centre of rotation (3, 2).

2 **(a)** See the diagram opposite.

(b) See the diagram opposite.

(a)

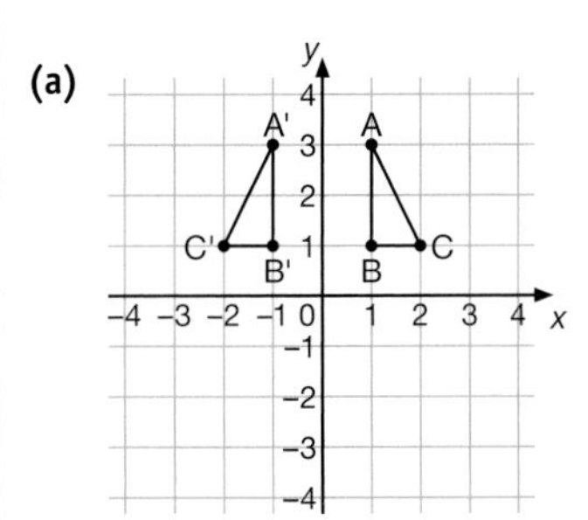

(b)

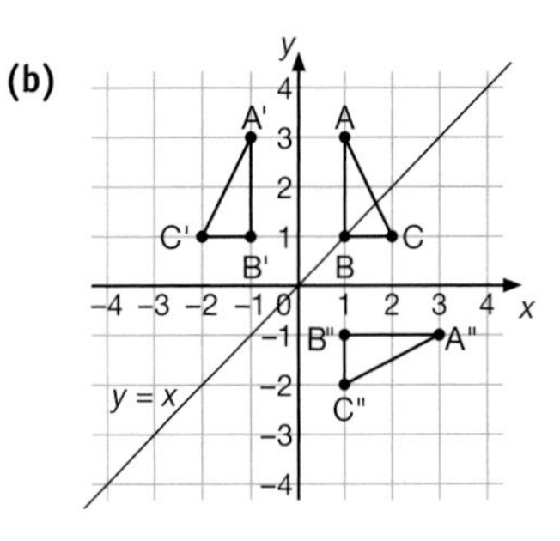

(c) Rotation of 90° anticlockwise about the origin.

(c) The question asks for a single transformation, and this is a rotation of 90° anticlockwise (or counterclockwise), centre of rotation (0, 0), or the origin.

3 **(a)** See the diagram opposite.

(a)

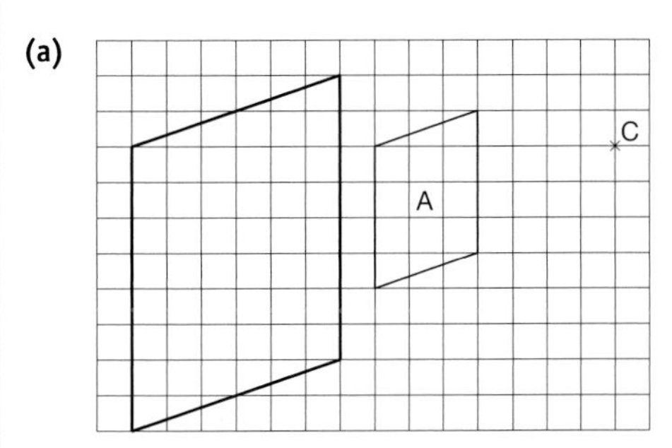

(b) **(i)** Area of A = 12 square units

(ii) Area of enlargement = 48 square units

(iii) 1 : 4

(b) **(i)** You can either count the squares or use the formula (remember to state the units).

(ii) Again, you can count the squares or use the formula (remember to state the units).

(iii) Ratio = 12 : 48 = 1 : 4 (cancelling down)

Answer

4 See the diagram opposite.
Coordinates of the points:
$A_2 = (2, 0)$
$B_2 = (6, 0)$
$C_2 = (4, -3)$
$D_2 = (0, -3)$

5 **(a)** See the diagram opposite.
(b) See the diagram opposite.
(c) Translation through vector $\begin{pmatrix} 5 \\ -1 \end{pmatrix}$.

How to solve these questions

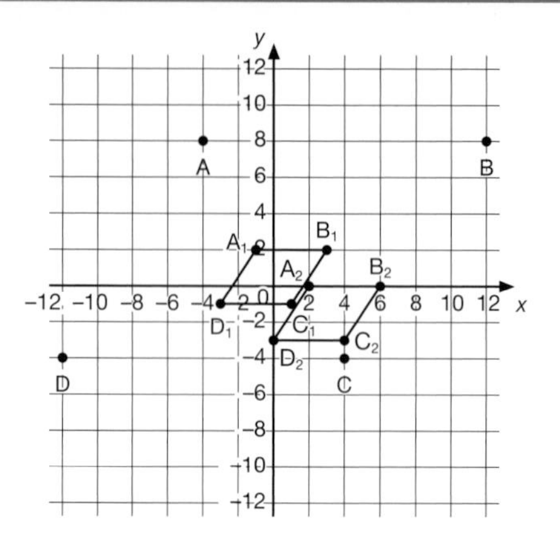

(a)

P Q

(b)

P Q R

(c) The single transformation is a translation. You should describe it fully, as a translation through vector $\begin{pmatrix} 5 \\ -1 \end{pmatrix}$.

Chapter 2 Pythagoras' theorem

Answer

1 **(a)** AC = 10.3 m (1 d.p.)

(b) AD = 6.6 m (1 d.p.)

How to solve these questions

(a)
$AC^2 = AB^2 + BC^2$
$AC^2 = 8.3^2 + 6.1^2$
$AC^2 = 68.89 + 37.21$
$AC^2 = 106.1$
$AC = \sqrt{106.1}$
$AC = 10.300\,485\,43$
$AC = 10.3$ m (1 d.p.)

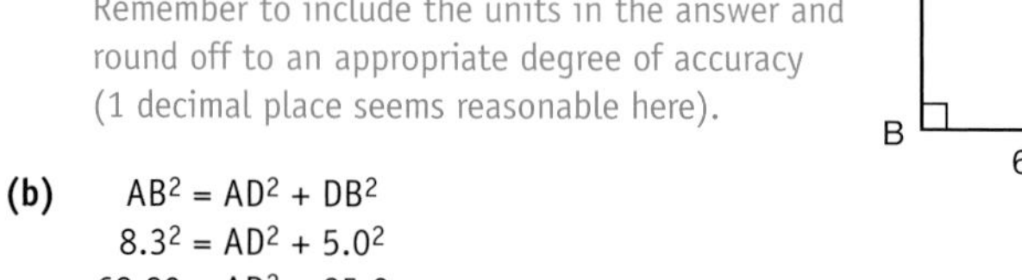

Remember to include the units in the answer and round off to an appropriate degree of accuracy (1 decimal place seems reasonable here).

(b)
$AB^2 = AD^2 + DB^2$
$8.3^2 = AD^2 + 5.0^2$
$68.89 = AD^2 + 25.0$
$AD^2 = 68.89 - 25.0$
$AD^2 = 43.89$
$AD = \sqrt{43.89}$
$AD = 6.624\,952\,83$
$AD = 6.6$ m (1 d.p.)

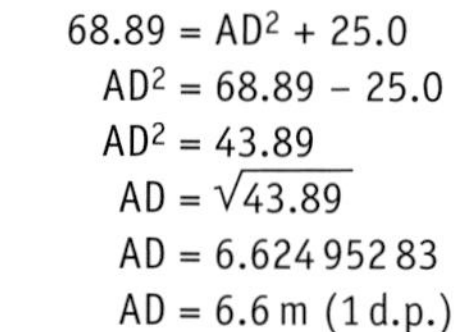

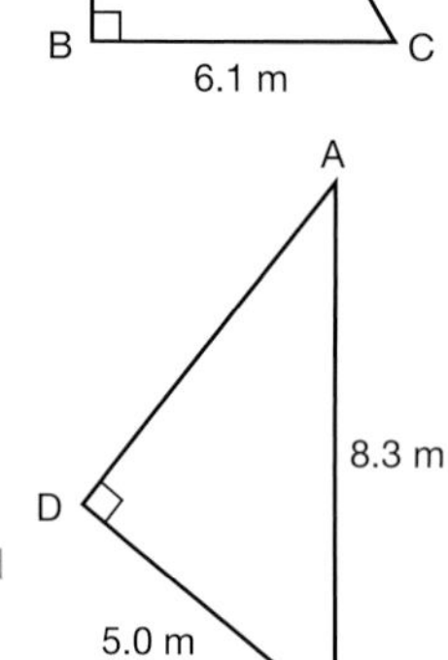

Remember to include the units in the answer and round off to an appropriate degree of accuracy (1 decimal place seems reasonable here).

Answer / How to solve these questions

2 The length of a diagonal $= \sqrt{200}$ cm

Using Pythagoras' theorem:

$x^2 = 10^2 + 10^2$
$x^2 = 100 + 100$
$x^2 = 200$
$x = \sqrt{200}$ (leaving in surd form)

The length of a diagonal $= \sqrt{200}$ cm

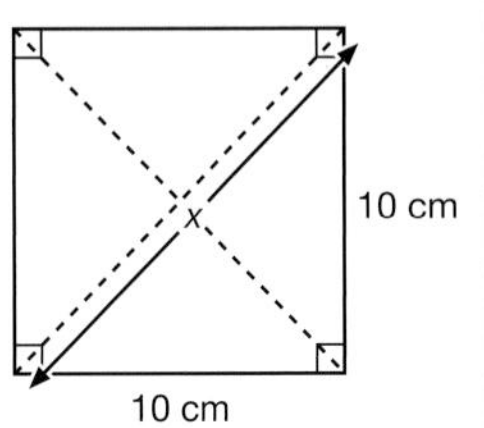

3 Height $= \sqrt{84}$ cm

To find the height, you can split the isosceles triangle into two right-angled triangles.

Use Pythagoras' theorem: $a^2 + b^2 = c^2$

$4^2 + h^2 = 10^2$
$16 + h^2 = 100$
$h^2 = 100 - 16$
$h^2 = 84$
$h = \sqrt{84}$ (leaving in surd form)

Height $= \sqrt{84}$ cm

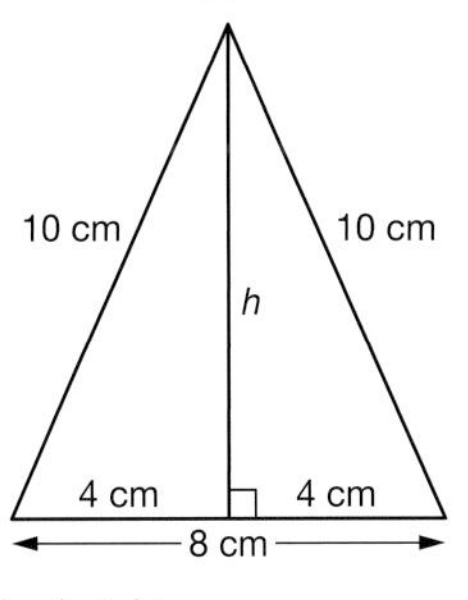

4 **(a)** Area $= 28\text{ cm}^2$

(b) Perimeter = 22.9 cm (1 d.p.)

(a) Area of the parallelogram = base × perpendicular height
$= (2 + 5) \times 4$
$= 28\text{ cm}^2$

(b) To work out the perimeter, you need to find the length of the slant height.

$x^2 = 4^2 + 2^2$
$x^2 = 16 + 4$
$x^2 = 20$
$x = \sqrt{20}$
$x = 4.472\,135\,955$

x
4
2
5
Not drawn to scale

Perimeter $= 7 + 4.472\,135\,955 + 7 + 4.472\,135\,955$
$= 22.944\,271\,91$
$= 22.9$ cm (1 d.p.)

5 $h = 2.08$ m (2 d.p.)

Since the triangle is right-angled, and you know two sides, then you can use Pythagoras' theorem to find the third side.

The length of the hypotenuse is given, so use:

$2.75^2 = h^2 + 1.80^2$
$7.5625 = h^2 + 3.24$
$h^2 = 7.5625 - 3.24$
$h^2 = 4.3225$
$h = 2.079\,062\,289$
$h = 2.08$ m (2 d.p.)

It is reasonable to correct the answer to two decimal places, as this is how the original numbers were given.

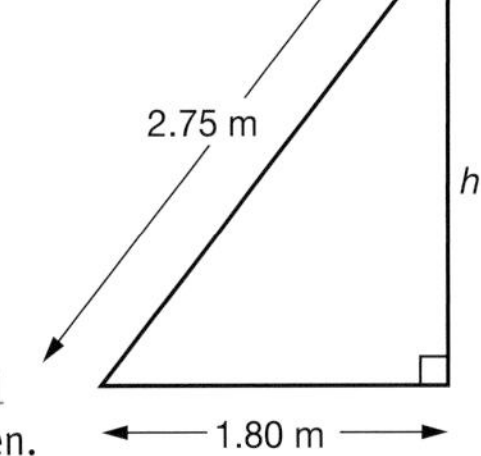

6 AB $= \sqrt{13}$ units

BC $= \sqrt{65}$ units

CA $= 2\sqrt{13}$ units

ABC is a right-angled triangle as $CA^2 + AB^2 = BC^2$

$AB^2 = 2^2 + 3^2$
$AB^2 = 4 + 9$
$AB = \sqrt{13}$ units

$BC^2 = 7^2 + 4^2$
$BC^2 = 49 + 16$
$BC = \sqrt{65}$ units

$AC^2 = 6^2 + 4^2$
$AC^2 = 36 + 16$
$AC = \sqrt{52}$ units

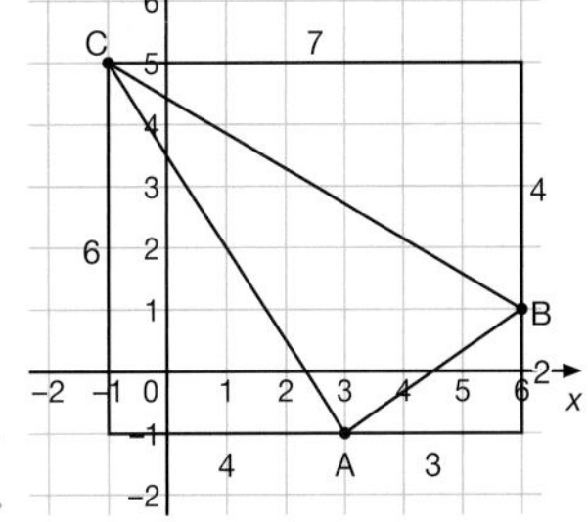

Hint: In examples like these it is usually better to leave the answers in surd (square root) form.

If ABC is a right-angled triangle then the square on the hypotenuse must be equal to the sum of the squares on the other two sides. (Pythagoras' theorem is true for all right-angled triangles.)

You need to show that $BC^2 = AC^2 + AB^2$

LHS $= BC^2 = (\sqrt{65})^2 = 65$

RHS $= AC^2 + AB^2 = (\sqrt{52})^2 + (\sqrt{13})^2$ as $(\sqrt{52})^2 = 52$ and $(\sqrt{13})^2 = 13$
$= 52 + 13 = 65$

As they are the same, this shows that the triangle ABC is right-angled.

Chapter 3 Angle and tangent properties of circles

Answer / How to solve these questions

1

(a) Angle ABC = 34°

(a) $\angle ACB = 90°$ (as the angle in a semicircle is always 90°)
$\angle ABC = 180° - 90° - 56°$
$= 34°$ (as the angles in triangle ABC total 180°)

(b) Angle AOD = 34°

(b) $\angle AOD = 34°$ (as corresponding angles between parallel lines OD and BC are equal, $\angle AOD = \angle ABC$)

(c) Angle ACD = 17°

(c) $\angle ACD = 17°$ (as the angle subtended by arc/chord AD at the centre is twice that subtended at the circumference)

(d) Angle OAD = 73°

(d) $\angle ODA = \angle OAD$ (base angles of isosceles triangle)
$\angle ODA + \angle OAD = 180° - 34°$ (angle sum of triangle)
$\angle ODA + \angle OAD = 146°$
$\angle OAD = \angle ODA = 73°$

(e) Angle BOT = 56°

(e) $\angle BOC = 112°$ (as the angle subtended by arc/chord BC at the centre is twice that subtended at the circumference)
$\angle BOT = \frac{1}{2} \times 112°$ (as $\angle COT = \angle BOT$ by the tangent properties of a circle)
$\angle BOT = 56°$

2

(a) $\angle POQ = 75°$

(a) $\angle POQ = 75°$ ($\angle POQ = \angle POR$ as triangles POQ and POR are congruent triangles)

(b) $\angle OPQ = 15°$

(b) $\angle OQP = 90°$ (as the tangent to a circle is perpendicular to the radius at the point of contact)
$\angle OPQ = 15°$ (as the angles of triangle OPQ total 180°)

(c) $\angle OPR = 15°$

(c) $\angle OPR = 15°$ ($\angle OPR = \angle OPQ$ as triangles POR and POQ are congruent triangles)

3 Angle POQ = 130°

$\angle CRO = \angle CQO = 90°$ (as the tangent to a circle is perpendicular to the radius at the point of contact)
$\angle RCO = \angle QCO = 20°$ ($\angle RCO = \angle QCO$ as the triangle RCO and triangle QCO are congruent triangles)
$\angle ROC = 70°$ (angle sum of triangle ROC)
$\angle QOC = 70°$ (angle sum of triangle QOC)
$\angle ROQ = 140°$ ($\angle ROQ = \angle ROC + \angle QOC$)
$\angle ROP = 90°$ (as the interior angles between parallel lines CA and OP are supplementary and add up to 180°)
$\angle POQ = 130°$ (as the angles in a circle add up to 360°)

4 The length of the chord is $2\sqrt{20}$ cm.

Let O be the centre of the circle and OX the perpendicular bisector of the chord.

Use Pythagoras' theorem in the right-angled triangle OAX.

$$6^2 = 4^2 + AX^2$$
$$AX^2 = 36 - 16$$
$$AX^2 = 20$$
$$AX = \sqrt{20}$$
$$AB = 2 \times AX = 2 \times \sqrt{20}$$
$$AB = 2 \times \sqrt{20} \text{ cm}$$
$$AB = 2\sqrt{20} \text{ cm}$$

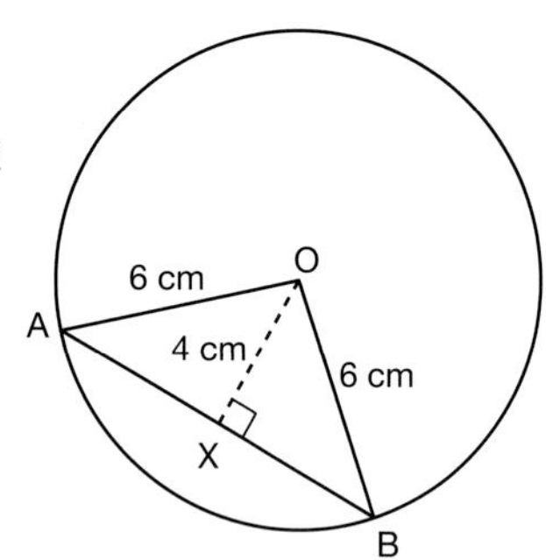

Answer

5 The distance between the chords is 27.9 cm (3 s.f.).

How to solve these questions

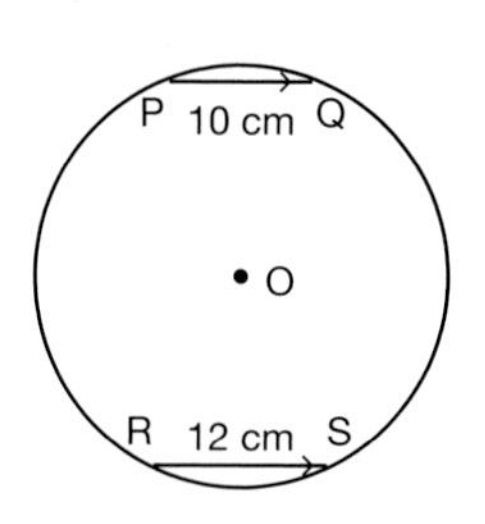

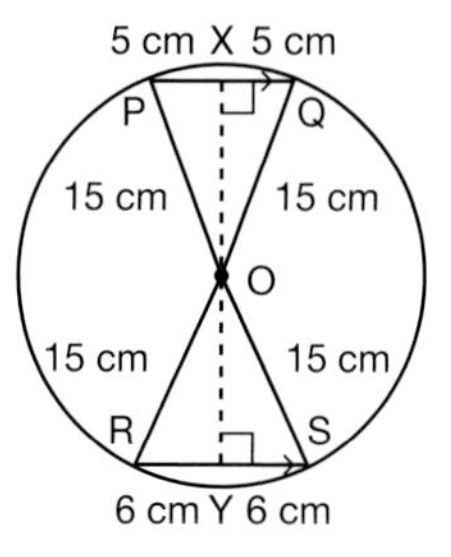

Use Pythagoras' theorem in the right-angled triangle ORY.

$15^2 = 6^2 + OY^2$

$225 = 36 + OY^2$

$OY^2 = 189$

$OY = 13.747\,727\,08$

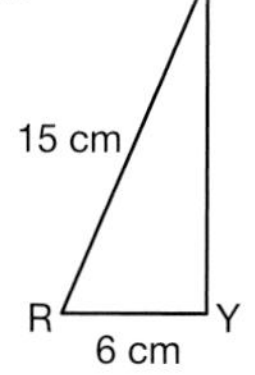

Now use Pythagoras' theorem in the right-angled triangle OPX.

$15^2 = 5^2 + OX^2$

$225 = 25 + OX^2$

$OX^2 = 200$

$OX = 14.142\,135\,62$

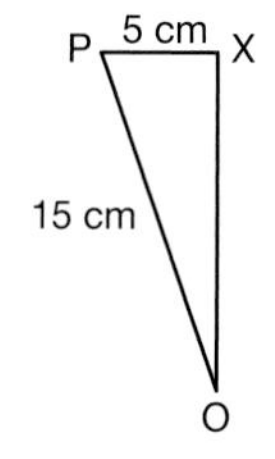

The distance between the two chords = OY + OX.

$OY + OX = 13.747\,727\,08 + 14.142\,135\,62$

$= 27.889\,862\,71$

$= 27.9$ cm (3 s.f.)

Chapter 4 Sine, cosine, tangent

Answer

1 Area of parallelogram = 37.2 cm^2 (3 s.f.)

How to solve these questions

First you need to find the height of the parallelogram.

The identified sides are 'opposite' and 'hypotenuse' so use:

$$\text{sine } x = \frac{\text{opposite}}{\text{hypotenuse}}$$

$$\sin 48° = \frac{h}{5}$$

$h = 5 \times \sin 48°$

$h = 5 \times 0.7431$

$h = 3.7155$

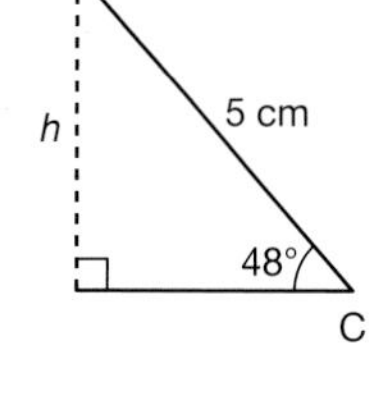

Area of parallelogram = base × perpendicular height

$= 10 \times 3.7155 = 37.155 = 37.2$ cm^2 (3 s.f.)

2 The required angle is 28.6° (3 s.f.)

The identified sides are 'adjacent' and 'opposite' so use:

$$\text{tangent } x = \frac{\text{opposite}}{\text{adjacent}}$$

$\tan x = \frac{6}{11}$

$x = \tan^{-1} \frac{6}{11}$

$x = 28.610\,459\,67°$

$x = 28.6°$ (3 s.f.)

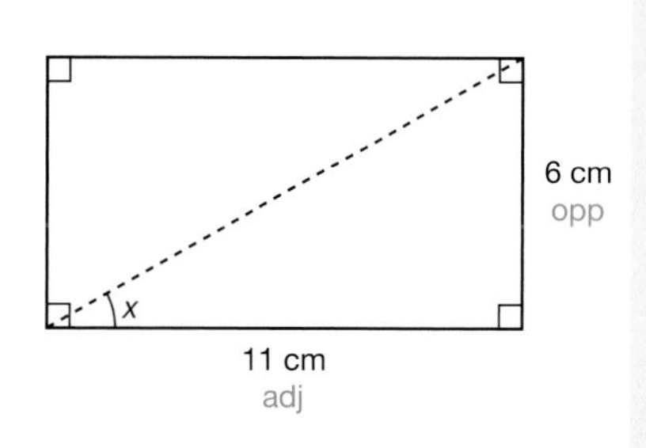

Answer / How to solve these questions

3 Distance north = 2.18 km (3 s.f.)

Take PS as the distance north.

$$\cos 85° = \frac{PS}{25}$$

$PS = 25 \times \cos 85°$

$PS = 2.178\,893\,569$ km

Distance north = 2.18 km (3 s.f.)

Distance east = 24.9 km (3 s.f.)

Take SR as the distance east.

$$\sin 85° = \frac{SR}{25}$$

$SR = 25 \times \sin 85°$

$SR = 24.904\,867\,45$ km

Distance east = 24.9 km (3 s.f.)

4 (a) The distance of A from B = 6.24 km (3 s.f.)

(a) You can use Pythagoras' theorem to calculate the third side.

$8^2 = AB^2 + 5^2$

$64 = AB^2 + 25$

$AB^2 = 64 - 25$

$AB^2 = 39$

$AB = 6.244\,997\,998$ km

$AB = 6.24$ km (3 s.f.)

(b) (i) $x = 51.3°$ (1 d.p.)

(b) (i) $\cos x = \frac{5}{8}$

$x = \cos^{-1} \frac{5}{8}$

$x = 51.317\,812\,55°$

$x = 51.3°$ (1 d.p.)

(ii) Bearing = 321.3° (1 d.p.)

(ii) Bearing

$= 270° + 51.317\,812\,55°$

$= 321.317\,812\,6°$

$= 321.3°$ (1 d.p.)

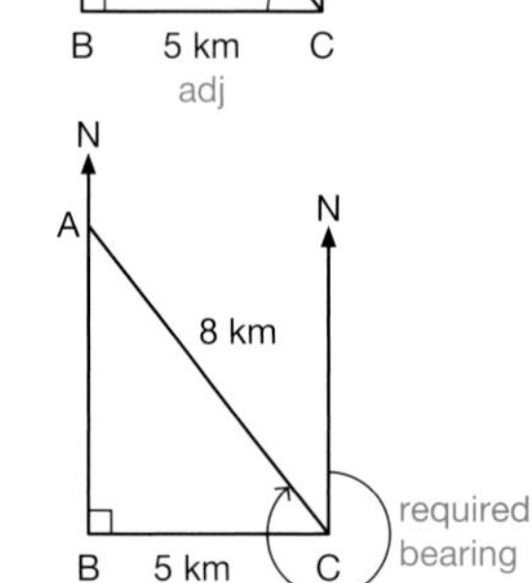

(iii) Bearing = 141.3° (1 d.p.)

(iii) $\angle BAC$

$= 90° - 51.317\,812\,55°$

$= 38.682\,187\,45°$

Bearing

$= 180° - 38.682\,187\,45°$

$= 141.317\,812\,5°$

$= 141.3°$ (1 d.p.)

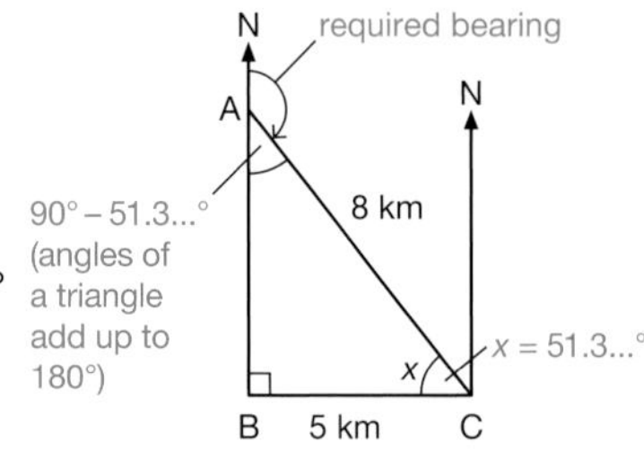

5 (a) Height of the cliff = 36.0 m (3 s.f.)

(a) $\tan 42° = \frac{h}{40}$

$h = 40 \times \tan 42°$

$h = 36.016\,161\,77$ m

$h = 36.0$ m (3 s.f.)

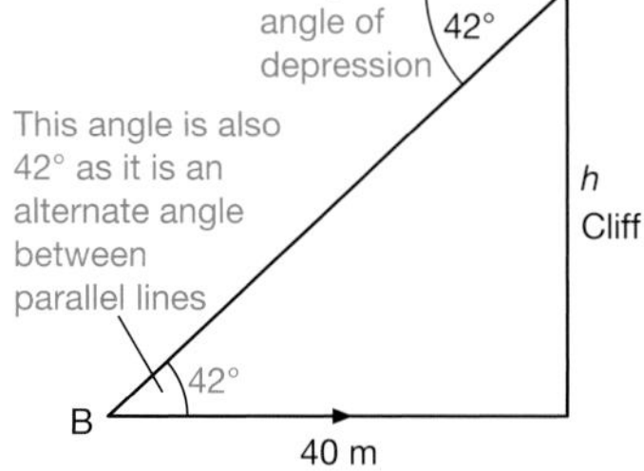

(b) Angle of elevation = 7.08° (3 s.f.)

(b) $\tan \theta = \frac{36.016\,161\,77}{290}$

$\theta = \tan^{-1} \frac{36.016\,161\,77}{290}$

$\theta = 7.079\,522\,804°$

$\theta = 7.08°$ (3 s.f.)

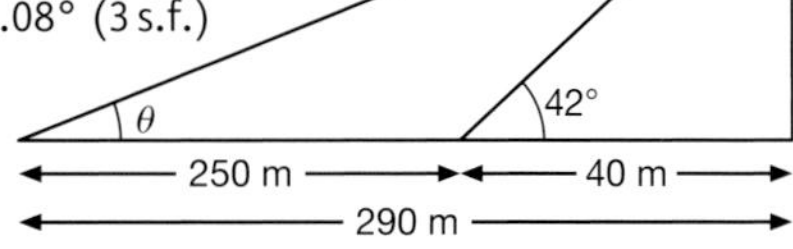

Chapter 5 Sine and cosine rules

Answer

1 $\cos A = -0.5$

So angle A is obtuse.

$A = 120°$

How to solve these questions

Draw a sketch so you can see what the question requires.

You are given three sides and have to find an angle, so you can use the cosine rule.

$$\cos A = \frac{b^2 + c^2 - a^2}{2bc}$$

Remember that the largest angle is always opposite the longest side.

$$\cos A = \frac{1.5^2 + 2.5^2 - 3.5^2}{2 \times 1.5 \times 2.5}$$

$$\cos A = \frac{2.25 + 6.25 - 12.25}{7.5} = \frac{-3.75}{7.5}$$

If the cosine of an angle is negative, the angle is obtuse, since the cosine of an obtuse angle is negative!

$\cos A = -0.5$

$A = 120°$

2 Angle ABC = 101.4° (1 d.p.)

You are given three sides and have to find an angle, so you can use the cosine rule.

$$\cos ABC = \frac{a^2 + c^2 - b^2}{2ac}$$

$$\cos ABC = \frac{3.1^2 + 8.6^2 - 9.7^2}{2 \times 3.1 \times 8.6}$$

$$\cos ABC = \frac{-10.52}{53.32}$$

$\cos ABC = -0.197\,299\,324$

The negative value means that angle ABC is obtuse.

$\angle ABC = 101.379\,075\,2°$

$\angle ABC = 101.4°$ (1 d.p.)

3 Distance = 26.3 km (3 s.f.)

Bearing = 046° (to the nearest whole degree)

Draw a sketch so you can see what the question requires.

From the sketch, the distance is represented by the line AC and the bearing is $32° + \angle BAC$.

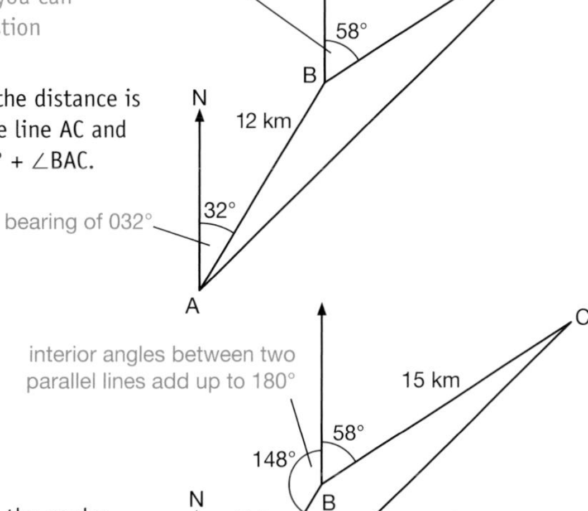

Add the details of the angles of triangle ABC to the sketch.

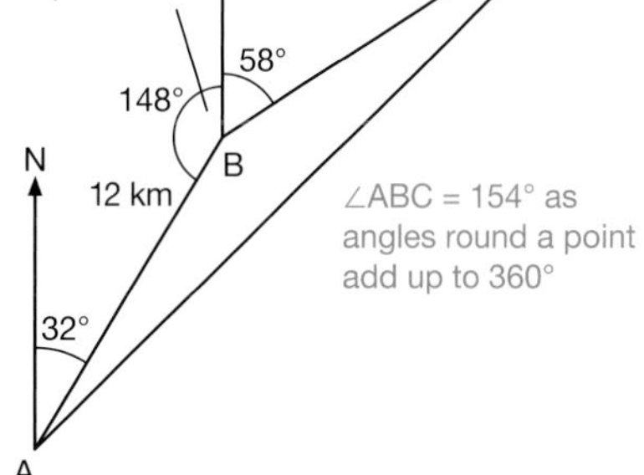

Answer

How to solve these questions

Now use the cosine rule on triangle ABC.

$b^2 = a^2 + c^2 - 2ac\cos B$

$AC^2 = 15^2 + 12^2 - (2 \times 15 \times 12 \times \cos 154°)$

$AC^2 = 369 - (-323.565\,856\,7)$

$AC^2 = 369 + 323.565\,856\,7$

$AC^2 = 692.565\,856\,7$

$AC = 26.316\,646$

So the distance = 26.3 km (3 s.f.).

The bearing is 32° + ∠BAC.

Use the sine rule to find angle BAC.

$$\frac{\sin A}{a} = \frac{\sin B}{b}$$

$$\frac{\sin BAC}{BC} = \frac{\sin ABC}{AC}$$

Substitute the given values, using the unrounded value for AC.

$$\frac{\sin BAC}{15} = \frac{\sin 154°}{26.316\,646}$$

$$\sin BAC = \frac{15 \times \sin 154°}{26.316\,646}$$

$\sin BAC = 0.249\,863\,421$

$\angle BAC = 14.469\,430\,3°$

The bearing = 32° + ∠BAC
= 32° + 14.469 430 3°
= 46.469 430 3°
= 046° (to the nearest whole degree)

4 Height of tower = 445m (3 s.f.)

∠ADC = 96° (as angles of triangle ADC add up to 180°)

Using the sine rule,

$$\frac{d}{\sin D} = \frac{a}{\sin A}$$

$$\frac{1000}{\sin 96°} = \frac{DC}{\sin 38°}$$

$$DC = \frac{1000\sin 38°}{\sin 96°}$$

$DC = 619.052\,710\,8$

$$\sin 46° = \frac{\text{opp}}{\text{hyp}}$$

$$\sin 46° = \frac{BD}{619.052\,710\,8}$$

$BD = 619.052\,710\,8 \times \sin 46°$

$BD = 445.309\,253\,4$

BD = 445 m (3 s.f.)

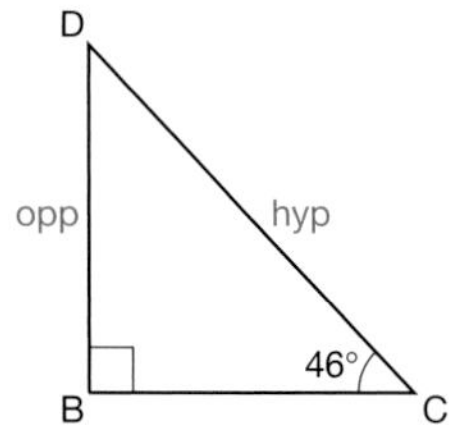

You can also find an answer using

$$\frac{1000}{\sin 96°} = \frac{DA}{\sin 46°}$$

Answer

5 AC = 6.05 cm (3 s.f.)

How to solve these questions

∠BAD = 55° (as the angles of the triangle BAD add up to 180°)

∠ADC = 75° (as the angles on a straight line add up to 180°)

Use the sine rule to find BD.

$$\frac{BD}{\sin BAD} = \frac{AB}{\sin ADB}$$

$$\frac{BD}{\sin 55°} = \frac{8.3}{\sin 105°}$$

$$BD = \frac{8.3 \times \sin 55°}{\sin 105°}$$

BD = 7.038 803 377

BC = BD + DC

BC = 13.138 803 38

Use the cosine rule on the triangle ABC.

$$b^2 = a^2 + c^2 - 2ac\cos B$$

$AC^2 = 13.138\,803\,38^2 + 8.3^2 - (2 \times 13.138\,803\,38 \times 8.3 \times \cos 20°)$

$AC^2 = 241.518\,154\,3 - (204.950\,847\,3)$

$AC^2 = 36.567\,306\,91$

$AC = 6.047\,090\,78$

So the distance AC = 6.05 cm (3 s.f.).

Chapter 6 Vectors

Answer

1 **(a)** $\mathbf{p} + \mathbf{q} = \begin{pmatrix} 7 \\ 1 \end{pmatrix}$

(b) $2\mathbf{p} - 3\mathbf{q} = \begin{pmatrix} 4 \\ 17 \end{pmatrix}$

(c) $|\mathbf{p} + \mathbf{q}| = \sqrt{50}$

2 **(a)** $\overrightarrow{PT} = \mathbf{a}$

(b) $\overrightarrow{PR} = \mathbf{b} + \mathbf{a}$

(c) $\overrightarrow{RT} = -\mathbf{b}$

(d) $\overrightarrow{SR} = \frac{3}{2}\mathbf{b}$

(e) $\overrightarrow{PS} = \mathbf{a} - \frac{1}{2}\mathbf{b}$

How to solve these questions

1

(a) $\mathbf{p} + \mathbf{q} = \begin{pmatrix} 5 \\ 4 \end{pmatrix} + \begin{pmatrix} 2 \\ -3 \end{pmatrix} = \begin{pmatrix} 7 \\ 1 \end{pmatrix}$

(b) $2\mathbf{p} - 3\mathbf{q} = 2 \times \begin{pmatrix} 5 \\ 4 \end{pmatrix} - 3 \times \begin{pmatrix} 2 \\ -3 \end{pmatrix} = \begin{pmatrix} 10 \\ 8 \end{pmatrix} - \begin{pmatrix} 6 \\ -9 \end{pmatrix} = \begin{pmatrix} 10 - 6 \\ 8 - -9 \end{pmatrix}$

$= \begin{pmatrix} 4 \\ 17 \end{pmatrix}$

(c) $|\mathbf{p} + \mathbf{q}| = 7^2 + 1^2$

$= \sqrt{50}$

2

(a) $\overrightarrow{PT} = \mathbf{a}$ (as PT is parallel to QR and the same length)

(b) $\overrightarrow{PR} = \overrightarrow{PQ} + \overrightarrow{QR}$

$= \mathbf{b} + \mathbf{a}$

(c) $\overrightarrow{RT} = -\mathbf{b}$ (as RT is parallel to QP and the same length)

(d) $\overrightarrow{SR} = \overrightarrow{ST} + \overrightarrow{TR}$

$= \frac{1}{2}\overrightarrow{TR} + \overrightarrow{TR} = \frac{3}{2}\overrightarrow{TR}$ (as TR is twice the length of ST)

$= \frac{3}{2}\mathbf{b}$

(e) $\overrightarrow{PS} = \overrightarrow{PT} + \overrightarrow{TS}$

$= \overrightarrow{PT} + \frac{1}{2}\overrightarrow{RT}$ (as TR is twice the length of ST)

$= \mathbf{a} + \frac{1}{2}(-\mathbf{b})$

$= \mathbf{a} - \frac{1}{2}\mathbf{b}$

Answer

3 **(a)** See the diagram opposite.

(b) $\overrightarrow{XP} = 2\frac{1}{2}\mathbf{a} + \frac{1}{2}\mathbf{b}$

How to solve these questions

(a)

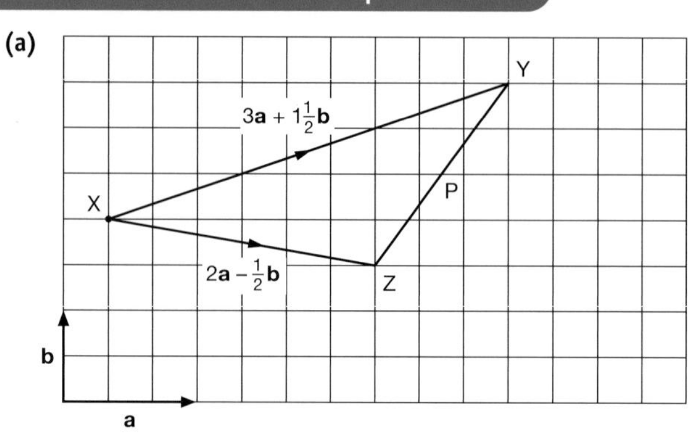

(b) $\overrightarrow{XP} = \overrightarrow{XZ} + \frac{1}{2}\overrightarrow{ZY}$

$= 2\mathbf{a} - \frac{1}{2}\mathbf{b} + \frac{1}{2}(\overrightarrow{ZX} + \overrightarrow{XY})$

$= 2\mathbf{a} - \frac{1}{2}\mathbf{b} + \frac{1}{2}\ (-2\mathbf{a} + \frac{1}{2}\mathbf{b} + 3\mathbf{a} + 1\frac{1}{2}\mathbf{b})$

$= 2\mathbf{a} - \frac{1}{2}\mathbf{b} - \mathbf{a} + \frac{1}{4}\mathbf{b} + 1\frac{1}{2}\mathbf{a} + \frac{3}{4}\mathbf{b}$

$= 2\frac{1}{2}\mathbf{a} + \frac{1}{2}\mathbf{b}$

Note that since you are told that you **'must** show all your working' then it is important to calculate the above rather than find it from the grid.

Answer

4 $\overrightarrow{XY} = \overrightarrow{XP} + \overrightarrow{PY}$

$XY = \mathbf{a} + \mathbf{b}$

But $\overrightarrow{QX} = \overrightarrow{XP}$

and $\overrightarrow{YR} = \overrightarrow{PY}$

$\overrightarrow{QR} = \overrightarrow{QP} + \overrightarrow{PR}$

$\overrightarrow{QR} = 2\mathbf{a} + 2\mathbf{b}$

$\overrightarrow{QR} = 2(\mathbf{a} + \mathbf{b})$

$\overrightarrow{QR} = 2\overrightarrow{XY}$

The magnitude of $\overrightarrow{QR}$ is twice the magnitude of $\overrightarrow{XY}$ so the length QR = 2XY.

As $\overrightarrow{QR}$ is a multiple of $\overrightarrow{XY}$, $\overrightarrow{QR}$ and $\overrightarrow{XY}$ are in the same direction and therefore are parallel.

How to solve these questions

From the diagram:

$\overrightarrow{XY} = \overrightarrow{XP} + \overrightarrow{PY}$

$\overrightarrow{XY} = \mathbf{a} + \mathbf{b}$

As X is the midpoint of QP then QX = XP and similarly as Y is the midpoint of PR then YR = PY.

From the diagram:

$\overrightarrow{QR} = \overrightarrow{QP} + \overrightarrow{PR}$

$\overrightarrow{QR} = 2\mathbf{a} + 2\mathbf{b}$

$\overrightarrow{QR} = 2(\mathbf{a} + \mathbf{b})$

$\overrightarrow{QR} = 2\overrightarrow{XY}$

It is important to give reasons for your conclusions.

- Since the magnitude of $\overrightarrow{QR}$ is twice the magnitude of XY then the length QR = 2XY.
- Similarly $\overrightarrow{QR}$ is a multiple of $\overrightarrow{XY}$ so $\overrightarrow{QR}$ and $\overrightarrow{XY}$ are in the same direction and parallel.

Answer

5 **(a)** $\overrightarrow{AC} = \mathbf{a} + \mathbf{b}$

(b) $\overrightarrow{AO} = \mathbf{b}$

(c) $\overrightarrow{OB} = \mathbf{a} - \mathbf{b}$

(d) $\overrightarrow{AD} = 2\mathbf{b}$

$\overrightarrow{AC} = \mathbf{a} + \mathbf{b}$ and $\overrightarrow{FD} = \mathbf{a} + \mathbf{b}$

$\overrightarrow{AF} = \mathbf{b} - \mathbf{a}$ and $\overrightarrow{CD} = \mathbf{b} - \mathbf{a}$

AC is parallel and equal to FD.

AF is parallel and equal to CD.

Therefore ACDF is a parallelogram.

How to solve these questions

(a) $\overrightarrow{AC} = \overrightarrow{AB} + \overrightarrow{BC} = \mathbf{a} + \mathbf{b}$

(b) $\overrightarrow{AO} = \mathbf{b}$

(c) $\overrightarrow{OB} = \overrightarrow{OA} + \overrightarrow{AB} = -\mathbf{b} + \mathbf{a} = \mathbf{a} - \mathbf{b}$

(d) $\overrightarrow{AD} = 2\mathbf{b}$

You must show your working and your reasons.

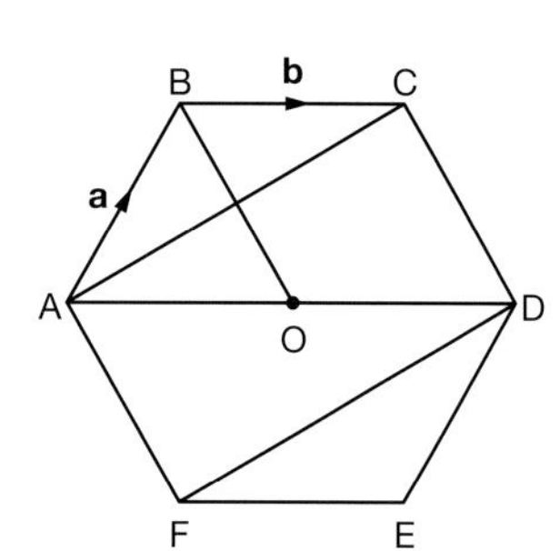

Chapter 7 Indices and standard form

Answer

1 $6 \times 2^4 = 6 \times 2 \times 2 \times 2 \times 2$
$= 96$

How to solve these questions

A common error here is to say $6 \times 2^4 = 12^4$ ✗
You must calculate 6×2^4 as $6 \times (2 \times 2 \times 2 \times 2) = 6 \times 16 = 96$

2 **(a)** 64

(b) 5

(c) 20

(a) $(2^3)^2 = (2 \times 2 \times 2)^2 = 8^2 = 64$
Also $(2^3)^2 = 2^3 \times 2^3 = 2^{3+3} = 2^6 = 64$

(b) $(\sqrt{5})^2 = \sqrt{5} \times \sqrt{5} = 5$
In general $(\sqrt{a})^2 = a$ always.

(c) $\sqrt{2^4 \times 25} = \sqrt{16 \times 25} = \sqrt{400} = 20$
An answer of -20 is also acceptable depending on the context.
Another method is:
$\sqrt{2^4 \times 25} = \sqrt{2^4 \times 5^2} = \sqrt{2^4} \times \sqrt{5^2} = 2^2 \times 5 = 20$
remembering that $\sqrt{a \times b} = \sqrt{a} \times \sqrt{b}$.

3 **(a)** $2.021\,22 \times 10^{-6}$
(b) $6.295\,302\,013 \times 10^{-11}$

You will be allowed to use a calculator for these questions, so practise keying in standard form numbers (use the EXP or EE keys).

2.02122 -6 or 2.02122 $^{-6}$

6.295302013 -11 or 6.295302013 $^{-11}$

Remember to write the results out in proper standard form.
It is always a good idea to check your answer by approximating.

4 **(a)** $pq = (6 \times 10^{-2}) \times (3 \times 10^{-4})$
$= 6 \times 10^{-2} \times 3 \times 10^{-4}$
$= 6 \times 3 \times 10^{-2} \times 10^{-4}$
$= 18 \times 10^{-2-4}$
$= 18 \times 10^{-6}$
$= 1.8 \times 10^{1} \times 10^{-6}$
$= 1.8 \times 10^{1-6}$
$= 1.8 \times 10^{-5}$

(a) Remember that pq means $p \times q$.
You can use the laws of indices ($a^m \times a^n = a^{m+n}$) to multiply 10^{-2} and 10^{-4}.
Remember that 18×10^{-6} is not in standard form.
You can use the fact that $18 = 1.8 \times 10$ or 1.8×10^1, with the laws of indices again, to find the correct answer.

(b) $\frac{p}{q} = \frac{6 \times 10^{-2}}{3 \times 10^{-4}}$
$= \frac{6}{3} \times \frac{10^{-2}}{10^{-4}}$
$= 2 \times 10^{-2--4}$
$= 2 \times 10^{-2+4}$
$= 2 \times 10^{2}$

(b) You can use the laws of indices ($a^m \div a^n = a^{m-n}$) to divide 10^{-2} by 10^{-4}.
Be careful with your signs on this equation.
$2 \times 10^2 = 200$ but the question says to leave your answer in standard form.

(c) $p - q = (6 \times 10^{-2}) - (3 \times 10^{-4})$
$= 0.06 - 0.0003$
$= 0.0597$
$= 5.97 \times 10^{-2}$

(c) Remember that numbers in standard form can only be added or subtracted if the power of 10 is the same in both of them.
Here it is best to convert the numbers to ordinary form and then subtract.
The answer can be given in standard form or left as 0.0597.
In this case, the question does not say which form to use, but you should always check.

5 8 minutes (to the nearest minute)

Remember that time $= \frac{\text{distance}}{\text{speed}}$
$= \frac{1.496 \times 10^{11}}{2.998 \times 10^{8}}$ seconds
$= 498.999\,332\,9$ seconds
$= 8.316\,655\,548$ minutes

Remember that the question says, 'Give your answer to the nearest minute' … don't lose those well-earned marks by not rounding your answer.
To convert seconds to minutes, you simply divide by 60.

Chapter 8 Percentages

Answer / How to solve these questions

1

$\frac{5}{9}$ = 55.555555...%

0.7 = 70%

63% = 63%

Least = $\frac{5}{9}$

Greatest = 0.7

In questions of this type, it is best to convert fractions and decimals to percentages.

To convert a fraction to a percentage you just multiply it by 100.

To convert a decimal to a percentage you just multiply it by 100.

2

Deposit = 5% of £740
= £37

Payments = 12 × £65
= £780

Total = £37 + £780
= £817

Difference = £817 − £740
= £77

For the deposit, find 10% of the cash price (= £74) then halve it to find 5%.

Calculate the total cost of twelve monthly payments of £65 (£780).

Add this to the deposit, to find the total credit price.

3

(a) 1.14 × 35 000 = 39 900

(b) 1.14 × 39 900 = 45 486

Remember that an increase of 14% gives a multiplier of $\frac{114}{100}$ = 1.14.

Alternatively, if you prefer not to use the multiplier method, you can work it out the long way.

4

£12 150

Remember that 10% represents $\frac{1}{10}$ so

$\frac{1}{10}$ of £15 000 = £1 500

After first year value = £15 000 − £1 500 = £13 500

$\frac{1}{10}$ of £13 500 = £1 350

After second year value = £13 500 − £1 350 = £12 150

Using 10% = $\frac{1}{10}$ is better than using a multiplier of 0.9 on a non-calculator question.

5

Percentage error = $\frac{0.005}{25.005} \times 100$

= 0.019 996%

You can use:

percentage error = $\frac{\text{error}}{\text{original amount}} \times 100$

where the error = 25.005 − 25 = 0.005 kg

and the original amount is the actual weight.

The calculator display of [1.9996 -04] or [1.9996 -04] needs to be interpreted.

6

88% = £550

1% = $\frac{£550}{88}$

= £6.25

100% = 100 × £6.25

= £625

Remember that £550 represents 100% − 12% = 88% of the original price.

Then 88% of the original price = £550

and 1% of the original price = $\frac{£550}{88}$ = £6.25

so 100% of the original price = 100 × £6.25 = £625

You should always make sure that your answer is reasonable – check it by reducing the price of £625 by 12% (equivalent to a multiplier of 0.88).

Chapter 9 Rational and irrational numbers

Answer / How to solve these questions

1 (a) 6.5

(a) $\sqrt{42} = 6.480\,706\,98\ldots$ $\sqrt{44} = 6.633\,249\,581\ldots$
so 6.5 ($= \frac{13}{2}$) is in the range, as are 6.49, 6.51, … and 6.495, 6.505, …
There are countless solutions. It is best to choose one of the most obvious.

(b) 2π

(b) Did you see that 2π (approximately 6.28) satisfies the inequality $6 < y < 7$?
You could also have suggested $\sqrt{37}$, $\sqrt{38}$, $\sqrt{39}$, $\sqrt{40}$, $\sqrt{41}$, $\sqrt{42}$, $\sqrt{43}$, $\sqrt{44}$, $\sqrt{45}$, $\sqrt{46}$, $\sqrt{47}$, $\sqrt{48}$ …

(c) $\frac{1}{3}$

(c) You could have written $\frac{1}{3}$, $\frac{1}{6}$, $\frac{1}{7}$, $\frac{1}{9}$, $\frac{1}{11}$, $\frac{1}{12}$, … in fact any fraction which produces a recurring decimal (see Chapter 10 *Recurring decimals*).

2 (a) $\sqrt{98} = 7\sqrt{2}$
(b) $\sqrt{5} + \sqrt{45} = 4\sqrt{5}$
(c) $\sqrt{6} \times \sqrt{12} = 6\sqrt{2}$

(a) $\sqrt{98} = \sqrt{49 \times 2} = \sqrt{49} \times \sqrt{2} = 7 \times \sqrt{2} = 7\sqrt{2}$
(b) $\sqrt{5} + \sqrt{45} = \sqrt{5} + \sqrt{9 \times 5} = \sqrt{5} + \sqrt{9} \times \sqrt{5} = \sqrt{5} + 3 \times \sqrt{5} = 4\sqrt{5}$
(c) $\sqrt{6} \times \sqrt{12} = \sqrt{6 \times 12} = \sqrt{72} = \sqrt{36 \times 2} = \sqrt{36} \times \sqrt{2} = 6 \times \sqrt{2} = 6\sqrt{2}$

3 $(1 + \sqrt{2})(1 - \sqrt{2})$
$= 1 \times 1 - 1 \times \sqrt{2} + \sqrt{2} \times 1 - \sqrt{2} \times \sqrt{2}$
$= 1 - \sqrt{2} + \sqrt{2} - 2$
$= -1$
so solution is rational

Remember that $\sqrt{2} \times \sqrt{2} = 2$.
-1 can be expressed as $\frac{-1}{1}$ so rational.

4 (a) $(3 - \sqrt{5})(4 + \sqrt{5})$
$= 12 + 3\sqrt{5} - 4\sqrt{5} - 5$
$= 7 - \sqrt{5}$
so $a = 7$

(a) $(3 - \sqrt{5})(4 + \sqrt{5}) = a - \sqrt{5}$
Start by expanding (multiplying out the brackets).
$(3 - \sqrt{5})(4 + \sqrt{5}) = 12 - 4\sqrt{5} + 3\sqrt{5} - 5 = 7 - \sqrt{5}$
Now compare the expressions on the right-hand side of the two equations.
$7 - \sqrt{5} = a - \sqrt{5}$ so $a = 7$.

(b) $(3 - \sqrt{5}) \times (4 + \sqrt{5}) + b$
$= 7 - \sqrt{5} + b$
so $b = \sqrt{5}$

(b) $(3 - \sqrt{5})(4 + \sqrt{5}) = 7 - \sqrt{5}$ (as before) so $(3 - \sqrt{5}) \times (4 + \sqrt{5}) + b = 7 - \sqrt{5} + b$
For this to be rational, you don't want any surds.
One possible solution is when b is equal to $\sqrt{5}$ as $7 - \sqrt{5} + \sqrt{5} = 7$ (which is rational).

5 (a) $pq - 1 = \sqrt{5} \times \sqrt{10} - 1$
$= \sqrt{50} - 1$
$= 5\sqrt{2} - 1$ (irrational)

(a) Substitute the values you are given into the expression.
The solution $\sqrt{50} - 1$ includes a surd, so it is irrational.

(b) $\frac{q}{pr} = \frac{\sqrt{10}}{\sqrt{5} \times \sqrt{8}} = \frac{\sqrt{10}}{\sqrt{40}}$
$= \sqrt{\frac{10}{40}} = \sqrt{\frac{1}{4}} = \frac{1}{2}$ (rational)

(b) Again, substitute the values you are given into the expression.
In the solution, all the surds cancel out, leaving a rational answer.

(c) $(p + q)^2 = (\sqrt{5} + \sqrt{10})^2$
$= 5 + \sqrt{50} + \sqrt{50} + 10$
$= 15 + 2\sqrt{50}$
$= 15 + 10\sqrt{2}$
(irrational)

(c) Substitute the values you are given into the expression and multiply out the brackets.
This leads to a solution which includes a surd, so the answer is irrational.

6 $5 + 2\sqrt{6}$

$(\sqrt{2} + \sqrt{3})^2 = (\sqrt{2} + \sqrt{3})(\sqrt{2} + \sqrt{3}) = (\sqrt{2})^2 + \sqrt{2} \times \sqrt{3} + \sqrt{3} \times \sqrt{2} + (\sqrt{3})^2$
$= 2 + 2\sqrt{2} \times \sqrt{3} + 3 = 5 + 2\sqrt{6}$
Remember that $\sqrt{a} \times \sqrt{b} = \sqrt{ab}$.

Chapter 10 Recurring decimals

Answer / How to solve these questions

1 **(a)** $0.\dot{4} = \frac{4}{9}$

(b) $0.\dot{5} = \frac{5}{9}$

$0.\dot{7} = \frac{7}{9}$

$0.\dot{8} = \frac{8}{9}$

Use the method described in the chapter, to answer this question.

What do you think $0.\dot{9}$ would be?

2 $0.\dot{2}\dot{1} = \frac{21}{99} = \frac{7}{33}$

$100 \times \text{(fraction)} = 21.2121212121\ldots$

and $1 \times \text{(fraction)} = 0.2121212121\ldots$

so $99 \times \text{(fraction)} = 21$ (subtracting)

and $\text{(fraction)} = \frac{21}{99}$ (dividing both sides by 99)

You must remember to cancel down or you will lose valuable marks.

Note: It is always a good idea to show all your working, just in case you make an error, so that the examiner can give you marks for the right method.

3 $0.\dot{3}\dot{2} = \frac{32}{99}$

so $201.\dot{3}\dot{2} = 201\frac{32}{99}$

Remember that a mixed number consists of a whole number plus a fraction.

Split up $201.\dot{3}\dot{2}$ into a whole number and a decimal fraction.

$201.\dot{3}\dot{2} = 201 + 0.\dot{3}\dot{2}$

Now change the decimal fraction into an ordinary fraction.

$100 \times \text{(fraction)} = 32.323232323232\ldots$

and $1 \times \text{(fraction)} = 0.323232323232\ldots$

so $99 \times \text{(fraction)} = 32$ (subtracting)

and $\text{(fraction)} = \frac{32}{99}$ (dividing both sides by 99)

so $0.\dot{3}\dot{2} = \frac{32}{99}$

Then $201.\dot{3}\dot{2} = 201 + \frac{32}{99} = 201\frac{32}{99}$

4 **(a)** $0.\dot{4}\dot{8} = \frac{48}{99} = \frac{16}{33}$

(b) $0.7\dot{4}\dot{8} = 0.7 + 0.0\dot{4}\dot{8}$

$= 0.7 + 0.\dot{4}\dot{8} \times \frac{1}{10}$

$= \frac{7}{10} + \frac{16}{33} \times \frac{1}{10}$

$= \frac{7}{10} + \frac{16}{330}$

$= \frac{231}{330} + \frac{16}{330}$

$= \frac{247}{330}$

(a) Use the method described in this chapter to answer this question.

(b) Split $0.7\dot{4}\dot{8}$ into decimal fractions that you can convert.

$0.7\dot{4}\dot{8} = 0.7 + 0.0\dot{4}\dot{8}$

$0.0\dot{4}\dot{8} = 0.\dot{4}\dot{8} \div 10$

Now use your answer in part (a).

Remember how to multiply and add fractions.

5 $\frac{1}{9} = 0.11111111\ldots$

$0.7\dot{1} = 0.7 + 0.01111111\ldots$

$= 0.7 + 0.11111111 \times \frac{1}{10}$

$= \frac{7}{10} + \frac{1}{9} \times \frac{1}{10}$

$= \frac{7}{10} + \frac{1}{90}$

$= \frac{63}{90} + \frac{1}{90}$

$= \frac{64}{90}$

$= \frac{32}{45}$

First, write down what you know.

$\frac{1}{9} = 0.11111111\ldots$

Now you need to use this fact in your answer.

Split $0.7\dot{1}$ into decimal fractions that you can convert.

$0.7\dot{1} = 0.7 + 0.01111111\ldots$

$0.01111111\ldots = 0.11111111\ldots \div 10$

Remember to cancel down and don't forget how to add and multiply fractions (you may need to revise this again).

Chapter 11 Upper and lower bounds

Answer

How to solve these questions

1

	Lower	Upper
(a)	4500	5500
(b)	4950	5050
(c)	3.615	3.625
(d)	235	245
(e)	$7\frac{1}{4}$	$7\frac{3}{4}$

Remember the rule and apply it, each time.

2 Shortest = 6694.5 km
= 6 694 500 m

Remember, the bounds are length $\pm \frac{1}{2}$ km (or $\pm$ 0.5 km).

3 Minimum volume
$= 11.5 \times 8.5 \times 9.5$
$= 928.625\ \text{cm}^3$

Maximum volume
$= 12.5 \times 9.5 \times 10.5$
$= 1246.875\ \text{cm}^3$

For a length of 12 cm the actual value could range from 11.5 cm to 12.5 cm.
Similarly, for a length of 9 cm the actual value could range from 8.5 cm to 9.5 cm and for a length of 10 cm the actual value could range from 9.5 cm to 10.5 cm.

The lower bound of the volume is found by multiplying all the minimum values.

The lower bound (minimum volume) $= 11.5 \times 8.5 \times 9.5$
$= 928.625\ \text{cm}^3$

The upper bound (maximum volume) $= 12.5 \times 9.5 \times 10.5$
$= 1246.875\ \text{cm}^3$

It is always best to use the same vocabulary as is used in the question so use 'lower and upper bounds' here.

4 **(a)** 18 years

(b) 18 years 364 days

(a) Other acceptable answers include
18 years 0 days or
18 years 1 second or
18 years 1 minute

(b) Other acceptable answers include
18 years 365 days (for a leap year)
18 years 11 months 30 (or 31) days
Note that 18.5 years is not an acceptable answer.

5 Smallest possible area of land
$= 28.5\% \times A_{min}$
$= \frac{28.5}{100} \times 4 \times \pi \times \left(\frac{12\,725}{2}\right)^2$
$= 144\,980\,740.9$
$= 1.449\,807\,409 \times 10^8$
$= 1.45 \times 10^8\ \text{km}^2$ (3 d.p.)

First, remember that the surface area of a sphere is given by $A = 4\pi r^2$ where r is the radius.

Notice also that this question involves the diameter (not the radius) so take care not to make mistakes.

$\text{percentage}_{min} = 28.5\%$ $\quad$ $\text{percentage}_{max} = 29.5\%$

$\text{diameter}_{min} = 12\,725$ km $\quad$ $\text{diameter}_{max} = 12\,775$ km

$\text{radius} = \frac{\text{diameter}}{2}$ so $\text{radius}_{min} = \frac{12\,725}{2} = 6362.5$ km

Remember to round your answer to an appropriate degree of accuracy.

6 Maximum speed $= \frac{65.5}{0.575}$
$= 113.913\,043\,5$
$= 114$ kph (3 s.f.)

Minimum speed $= \frac{64.5}{0.591\,666\,666}$
$= 109$ kph (3 s.f.)

First, find the minimum and maximum values:

$\text{distance}_{min} = 64.5$ km $\quad$ $\text{distance}_{max} = 65.5$ km

$\text{time}_{min} = 34.5$ minutes $\quad$ $\text{time}_{max} = 35.5$ minutes

Then use:

$\text{speed} = \frac{\text{distance}}{\text{time}}$

and remember that time must be expressed in hours.

34.5 minutes $= \frac{34.5}{60} = 0.575$ hours

35.5 minutes $= \frac{35.5}{60} = 0.591\,666\,666$ hours

So $\text{speed}_{max} = \frac{65.5}{0.575} = 113.913\,043\,5$ kph

And $\text{speed}_{min} = \frac{64.5}{0.591\,666\,666} = 109.014\,084\,5$ kph

Finally, round your answer to a reasonable degree of accuracy (say 3 s.f. here).

Chapter 12 Cumulative frequency

Answer

1 Range = 28 cm

Interquartile range = 23 cm

How to solve these questions

Start by arranging the information in order.

10, 12, 13, 13, 16, 18, 18, 19, 20, 20, 21, 36, 37, 37, 38

The range of the heights is 38 − 10 = 28 cm.

To find the interquartile range you need to find the lower quartile and the upper quartile. The lower quartile is the $\frac{1}{4}(15 + 1)$ = 4th value and the upper quartile is the $\frac{3}{4}(15 + 1)$ = 12th value.

From the data (all measured in cm):

10, 12, 13, 13, 16, 18, 18, 19, 20, 20, 21, 36, 37, 37, 38

Lower quartile (13) Median (19) Upper quartile (36)

Interquartile range = upper quartile − lower quartile = 36 − 13
= 23 cm

2

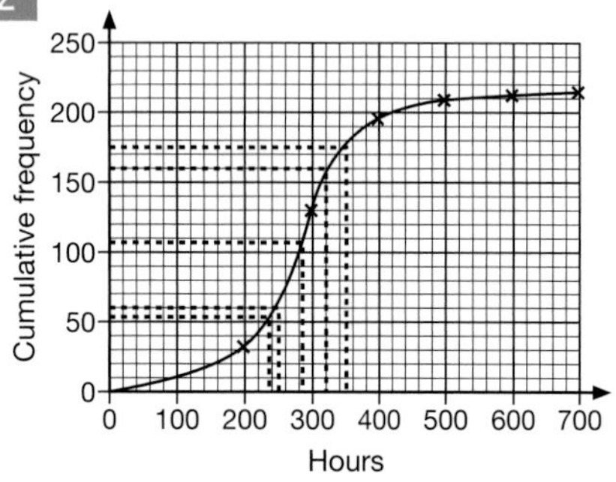

(a) 60 batteries lasted less than 250 hours.

(b) 39 batteries lasted more than 350 hours.

(c) Median = 282
Interquartile range = 100

Start by completing the cumulative frequency table.

Lifetime (hours)	Frequency	Cumulative frequency
100–199	32	32
200–299	98	130
300–399	65	195
400–499	14	209
500–599	3	212
600–699	2	214

Then draw the curve, as shown.

Note: The upper values of the class intervals are $199\frac{1}{2}$, $299\frac{1}{2}$, $399\frac{1}{2}$ as the lifetime is given to the nearest hour.

From the graph:

(a) 60 batteries lasted less than 250 hours

(b) 175 batteries lasted less than 350 hours
so 214 − 175 = 39 lasted more than 350 hours.

(c) median = 282 (read off value)
upper quartile = 320 (read off value)
lower quartile = 240 (read off value)
interquartile range = upper quartile − lower quartile
= 320 − 240
= 80

3 (a)

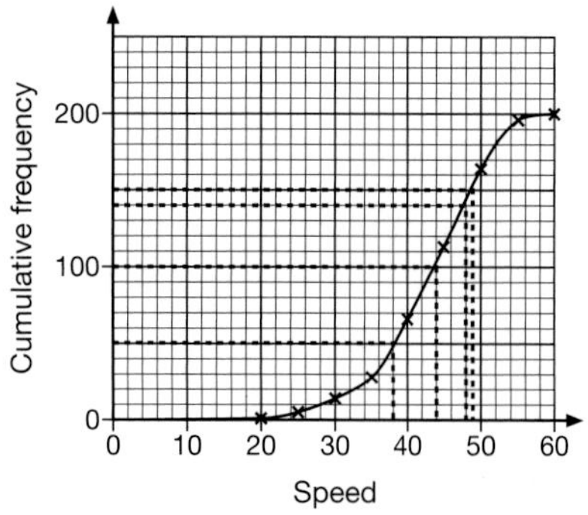

(a) Draw the curve from the table.

Note: The upper values of the class intervals are clearly 20, 25, 30, 35 … here.

Answer

How to solve these questions

(b) (i) Median = 44 mph

(ii) Interquartile range = 11 mph

(iii) Percentage of cars travelling at less than 48 mph = 70%

(b) (i) Median = 44 mph

(ii) LQ = 38
UQ = 49
Interquartile range = 49 − 38
= 11 mph

(iii) Number of cars travelling at less than 48 mph = 140

Percentage of cars = $\frac{140}{200} \times 100\%$
= 70%

4 (a) See the table opposite.

(a)

Time	$t < 10$	$t < 15$	$t < 20$	$t < 25$
Frequency	0	2	6	13
Time	$t < 30$	$t < 35$	$t < 40$	$t < 45$
Frequency	28	42	47	48

(b)

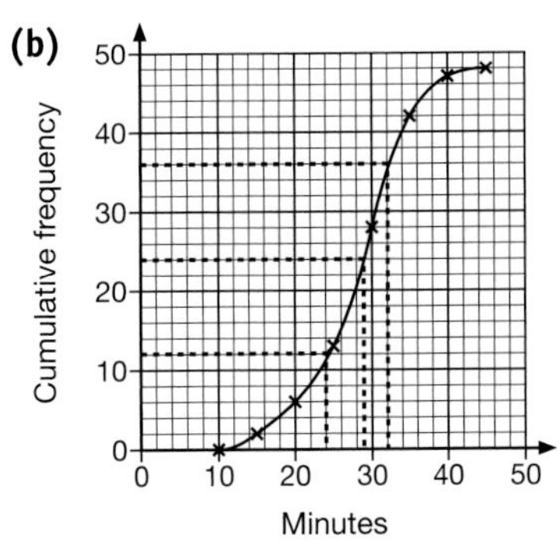

(b) Use the table to draw the curve.

(c) Median arrival time = 7.29 a.m.

(d) (i) Time of arrival = 7.24 a.m.

(ii) Petra waited 8 minutes.

(e) 12 people arrived after Karen.

(c) Calculate the median arrival time using a cumulative frequency value of 24. Median time of arrival = 7.29 a.m.

(d) (i) Find the time of arrival, by scanning across from the cumulative frequency of 12. Read the corresponding time (24 minutes), and add on to 7 a.m. Petra's time of arrival = 7.24 a.m.

(ii) Use a cumulative frequency value of $\frac{3}{4} \times 48 = 36$ to find when the coach arrived. This gives a time of 32 minutes, so the coach arrived at 7.32 a.m. Petra arrived at 7.24 a.m, so she waited 8 minutes.

(e) Now find the cumulative frequency at time = 32 minutes and subtract from 48, to find the number of people that arrived after Karen. Number of people who arrived after Karen = 12.

5 (a) Median height of the boys = 168 cm

(a)

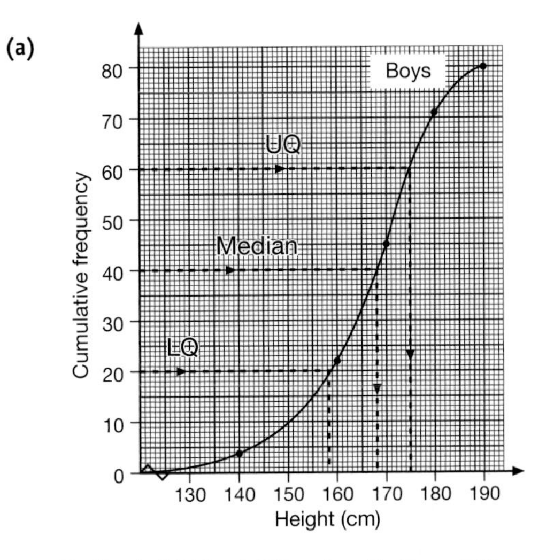

Median height is at 40th value. Median = 168 cm.

Answer

(b) Interquartile range = 17 cm

(c) (i) See the diagram opposite.

(ii) Boys have a bigger median height. Boys' heights are more spread out.

How to solve these questions

(b) Draw lines at the 20th and 60th values to calculate the interquartile range.
IQR = 175 – 158 = 17 cm

(c) (i)

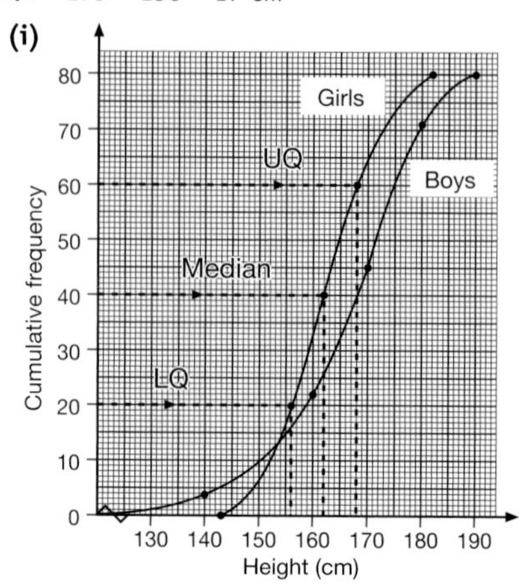

Chapter 13 Mode, median and mean

Answer

1 (a)

Time (t) seconds	Tally	Frequency
$0 \leq t < 5$	///	3
$5 \leq t < 10$	~~////~~ //	7
$10 \leq t < 15$	~~////~~ ////	9
$15 \leq t < 20$	////	4
$20 \leq t < 25$	//	2

(b) 10–15

2 (a) (i) Median runs per ball is 2.

(ii) Mean is 2.28.

(b) (i) New median is 2.5.

How to solve these questions

(a) The tallies are a useful way of recording the items of data. When you have finished, make sure that the total is the same as the number of entries in the table you started with.

Be careful to include 10.0 in the $10 \leqslant t < 15$ group.

(b) The modal class interval is the one with the largest frequency, which is $10 \leqslant t < 15$.

(a) (i) You need to look at the 13th value. The median number of runs per ball is 2.

(ii)

Runs	Frequency	Frequency × runs
x	f	fx
0	3	0
1	8	8
2	4	8
3	3	9
4	5	20
5	0	0
6	2	12
	$\sum f = 25$	$\sum fx = 57$

$$\text{Mean} = \frac{\sum fx}{\sum f} = \frac{57}{25} = 2.28$$

(b) (i) Add the new totals and amend the table.

Runs	Frequency	Frequency × runs
x	f	fx
0	3	0
1	8	8
2	4	8
3	4	12
4	7	28
5	1	5
6	3	18
	$\sum f = 30$	$\sum fx = 79$

Now you need to look at the $15\frac{1}{2}$th value.
The median of runs per ball is 2.5.

Answer	How to solve these questions
(ii) New mean = 2.63 (3 s.f.).	**(ii)** $\Sigma f = 25 + 5 = 30$ $\Sigma fx = 57 + 4 + 4 + 5 + 3 + 6 = 79$ Mean $= \frac{\Sigma fx}{\Sigma f} = \frac{79}{30} = 2.633\,333\,333$ $= 2.63$ (3 s.f.)
(c) The mean takes account of all the values.	**(c)** The mean takes account of all values and gives due weighting to the high number of 1s and 4s.
3 (a) (i) Range = 12 cm	**(a) (i)** Remember that the range should always be given as a single figure. The range of the lengths = 39 − 27 = 12 cm.
(ii) Mean length = 32 cm	**(ii)** Mean length $= \frac{27 + 28 + 29 + 30 + 31 + 31 + 32 + 33 + 35 + 37 + 39}{11}$ $= \frac{352}{11}$ = 32 cm
(b) The Spanish cucumbers have a smaller average length and their lengths are less spread out.	**(b)** The range is smaller for the Spanish cucumbers, so the lengths are not so spread out. The Spanish cucumbers also have a smaller average length.
4 (a) Range = 5 g	**(a)** The range of the weights in the table = 460 − 455 = 5 g
(b) Mean weight = 457 g (3 s.f.)	**(b)** See table below.

Weight (g)	Number of jars	
x	f	fx
454	0	0
455	1	455
456	6	2736
457	7	3199
458	3	1374
459	1	459
460	2	920
	$\Sigma f = 20$	$\Sigma fx = 9143$

For the frequency distribution, the mean $= \frac{\Sigma fx}{\Sigma f}$

$= \frac{9143}{20}$

$= 457.15$

= 457 g (3 s.f.)

5 Mean = 17 minutes (to the nearest minute)

Time, t (mins)	Number of people	Mid-interval value	Frequency × mid-interval value
$0 < t \le 10$	4	5	20
$10 < t \le 20$	7	15	105
$20 < t \le 30$	3	25	75
$30 < t \le 40$	2	35	70
	$\Sigma f = 16$		$\Sigma fx = 270$

Mean $= \frac{\Sigma fx}{\Sigma f}$

$= \frac{270}{16}$

$= 16.875$

= 17 minutes (to the nearest minute)

Answer

6 Mean height = 43 cm (2 s.f.)

7 **(a)** Mean height = 162 cm (3 s.f.)

(b) The median is the 30th value, and this value will be in the 156–164 cm group.

(c)

How to solve these questions

Height	Frequency	Mid-interval value	Frequency × mid-interval value
	f	x	$f \times x$
15–20	8	17.5	8 × 17.5 = 140
20–30	4	25	4 × 25 = 100
30–40	5	35	5 × 35 = 175
40–50	11	45	11 × 45 = 495
50–60	17	55	17 × 55 = 935
60–70	2	65	2 × 65 = 130
70–80	1	75	1 × 75 = 75
	$\Sigma f = 48$		$\Sigma fx = 2050$

For the group frequency distribution, mean = $\frac{\Sigma fx}{\Sigma f} = \frac{2050}{48}$

= 42.708 333 33

= 43 cm to an appropriate degree of accuracy

(a)

Height, h (cm)	Frequency	Mid-interval values	Frequency × mid-interval values
$140 \le h < 148$	6	144	864
$148 \le h < 156$	10	152	1520
$156 \le h < 164$	15	160	2400
$164 \le h < 172$	18	168	3024
$172 \le h < 180$	11	176	1936
	$\Sigma f = 60$		$\Sigma fx = 9744$

An estimate of the mean = $\frac{\Sigma fx}{\Sigma f}$

$= \frac{9744}{60}$

= 162.4

= 162 cm (3 s.f.)

(b) The median is the middle value (i.e. the 30th value) which lies in the $156 \leqslant h \leqslant 164$ group so that the median lies between 156 cm and 164 cm.

(c) Use the table to draw the histogram.

Since the class intervals are all equal, the histogram is similar to a barchart.

Chapter 14 Scatter diagrams

Answer

1

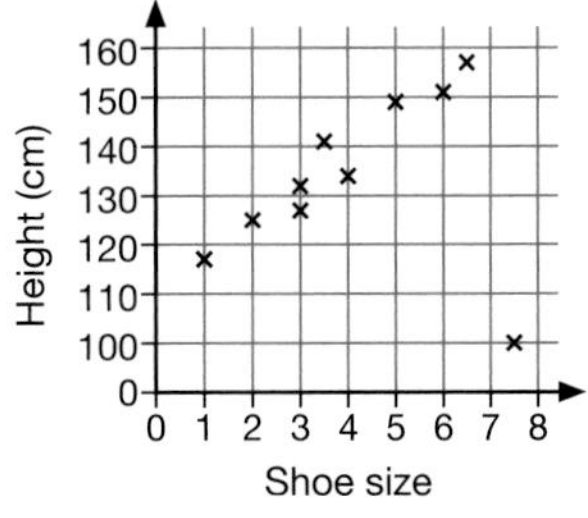

The relationship is strong and positive. The point ($7\frac{1}{2}$, 100) is likely to be an error.

2 **(a)**

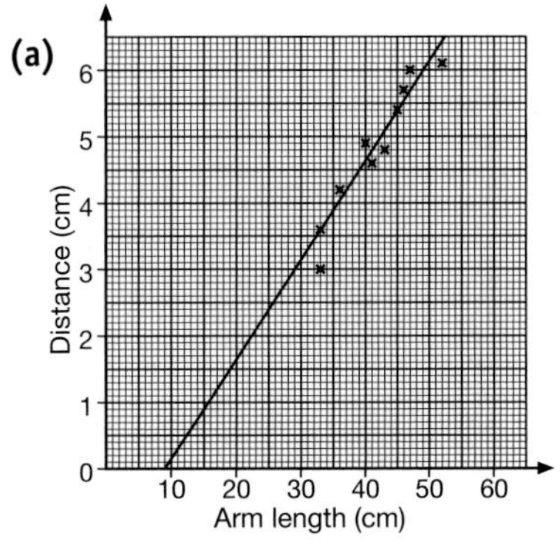

(b) **(i)** The distance thrown = 6.1 m

(ii) The arm length = 46 cm

3 **(a)** See the diagram opposite.

(b) **(i)** Height after 10 days = 11 cm

(ii) Height after 20 days = 21.5cm

(c) Answer (i) is more likely to be reliable because it is nearer the middle of the range, rather than close to one end.

How to solve these questions

Use the table to plot the information on a graph.

Look at the shape of the graph and the pattern formed by the dots to make a comment on the correlation.

(a) Plot the points and add the line of best fit, to pass through as many points as you can, but make sure that there are equal numbers of points on either side of the line.

(b) You can add lines to your graph, if it helps to find the answers.

(i) The distance thrown for an arm length of 50 cm = 6.1 m

(ii) The arm length for a distance of 5.5 m thrown = 46 cm

Look carefully at the points and try to draw the line of best fit through as many as possible, with equal numbers on either side.

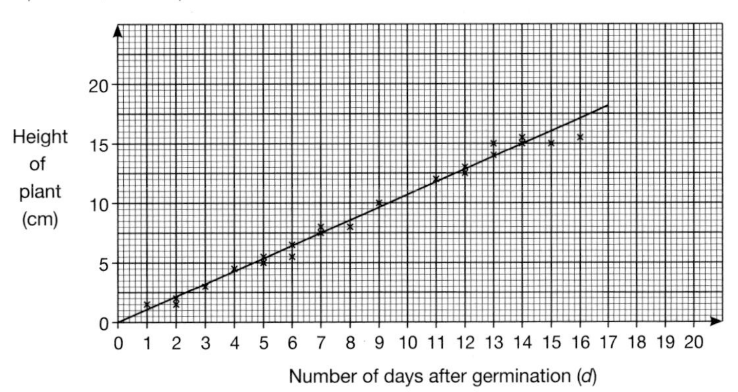

(b) Use the line of best fit to find:

(i) height after 10 days = 11 cm

(ii) height after 20 days = 21.5cm

(c) The answer which is more likely to be more reliable is **(i)**.

This is because it is in the middle of the range. 20 days is beyond one of the limits.

Answer

4 **(a)** See the diagram opposite.

(b) There seems to be strong positive correlation between adult literacy and life expectancy.

(c) Life expectancy of an adult from a country with 42% adult literacy = 52 years.

(d) Yes, it does seem reasonable since there seems to be positive correlation between the two sets of figures. It would also be reasonable to think that where adult literacy is high, the population would be able to read about healthy living.

How to solve these questions

(a) Use the figures to draw the scatter diagram.

(b) Look carefully at the graph and see if there is a pattern.
Describe any correlation that you can see between the 'Adult literacy' and the 'Life expectancy' figures.

(c) Draw a line of best fit.
From the graph, the life expectancy of an adult from a country with 42% adult literacy = 52 years.

(d) Think of a reason for your answer.

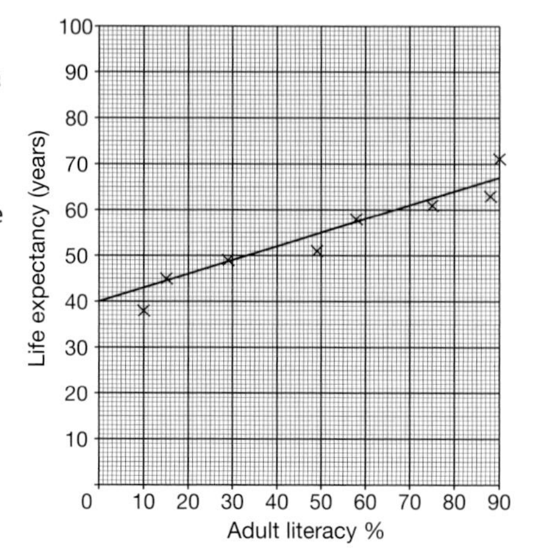

Chapter 15 Probability

Answer

1 **(a)** **(i)** Might be correct
(ii) Must be correct
(iii) Must be wrong

(b) **(i)** $\frac{1}{3}$
(ii) $\frac{7}{12}$
(iii) 0

2 **(a)** $\frac{3}{10}$

(b) $\frac{7}{10}$

3 **(a)** $\frac{2}{5}$

(b) 120

(c) $\frac{5}{12}$

How to solve these questions

(a) **(i)** The probability that the marble is red is $\frac{1}{2}$ might be correct.
(ii) The probability that the marble is red or blue is 1 must be correct.
(iii) The probability that the marble is red is −0.4 must be wrong.

(b) **(i)** p(blue) = $\frac{4}{12} = \frac{1}{3}$ (cancel wherever possible)
(ii) p(blue or red) = $\frac{7}{12}$
(iii) p(yellow) = 0 (impossible)

(a) p(a blue Ford is chosen from the cars) = $\frac{\text{number of blue Fords}}{\text{number of cars}}$
$= \frac{6}{20} = \frac{3}{10}$

(b) p(a Vauxhall is chosen from the white cars) = $\frac{\text{number of Vauxhalls}}{\text{number of white cars}}$
$= \frac{7}{10}$

(a) p(two occupants) = $\frac{20}{50} = \frac{2}{5}$

(b) Total number of car occupants $= 13 \times 1 + 20 \times 2 + 3 \times 3 + 12 \times 4 + 2 \times 5$
$= 13 + 40 + 9 + 48 + 10$
$= 120$

(c) p(driving a car) = $\frac{50}{120} = \frac{5}{12}$ (as there were 50 cars therefore 50 drivers)

Answer

4 **(a)** $\frac{3}{5}$

(b) See the diagram opposite.

(c) 500

5 **(a)**

Andy — Baljit

0.8 win: 0.6 win; 0.4 not win

0.2 not win: 0.6 win; 0.4 not win

(b) Probability = 0.44

6

(a) p(total is 10) = $\frac{1}{25}$

(b) p(total is 9) = $\frac{2}{25}$

(c) $\frac{3}{25}$

(d) $\frac{1}{125}$

How to solve these questions

4

(a) Relative frequency = $\frac{12}{20} = \frac{3}{5}$

(b) Use the data to draw the graph to show the relative frequency of throwing a head.

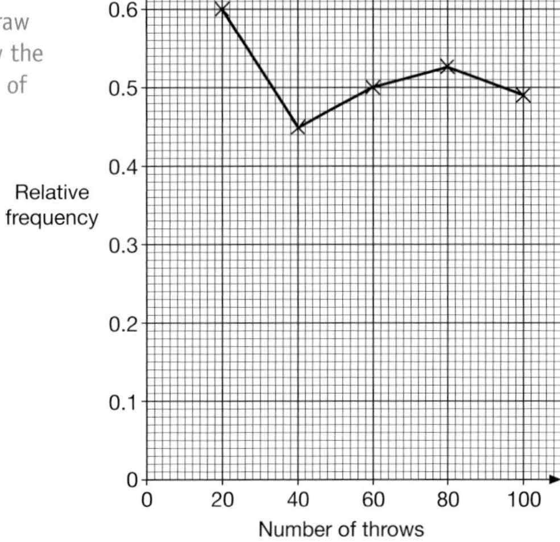

Number of throws	20	40	60	80	100
Number of heads	12	18	30	42	49
Relative frequency	$\frac{12}{20} = 0.6$	$\frac{18}{40} = 0.45$	$\frac{30}{60} = 0.5$	$\frac{42}{80} = 0.525$	$\frac{49}{100} = 0.49$

The relative frequency is tending towards 0.5.

(c) Relative frequency of throwing a head for 1000 throws
$= 1000 \times \frac{1}{2} = 500$

5

(a) Remember that the sum of the probabilities at the ends of the branches is 1.

(b) The probability that only one of the boys wins is the probability that Andy wins and Baljit does not or Andy does not win and Baljit does.

p(one boy wins) = p(Andy and not Baljit) + p(not Andy and Baljit)
$= 0.8 \times 0.4 + 0.2 \times 0.6$
$= 0.32 + 0.12$
$= 0.44$

6

Draw a space diagram to show the totals.

Totals

Spinner 2					
5	6	7	8	9	10
4	5	6	7	8	9
3	4	5	6	7	8
2	3	4	5	6	7
1	2	3	4	5	6
Spinner 1	1	2	3	4	5

(a) p(total is 10) = $\frac{1}{25}$

(b) p(total is 9) = $\frac{2}{25}$

(c) p(total is 3 or less) = p(total is 2 or total is 3)
= p(total is 2) + p(total is 3)
$= \frac{1}{25} + \frac{2}{25} = \frac{3}{25}$

(d) p(all three spinners show 5) = $\frac{1}{5} \times \frac{1}{5} \times \frac{1}{5} = \frac{1}{125}$

Answer

7 **(a)** (1, Heads), (1, Tails), (2, Heads), (2, Tails), (3, Heads), (3, Tails), (4, Heads), (4, Tails), (5, Heads), (5, Tails)

(b) **(i)** 0.14

(ii) 0

(iii) The spinner is most likely to land on 1.

(iv) 0.25

(c) 0.125

How to solve these questions

(a) Make sure that you list all of the possible outcomes. Be systematic, don't just write them down in any order.

(b) **(i)** p(spinner lands on 5) = 1 − (0.36 + 0.1 + 0.25 + 0.15) = 1 − 0.86 = 0.14

(ii) p(spinner lands on 6) = 0 (no chance of it landing on 6)

(iii) The spinner is most likely to land on 1. (This value has the highest probability so must be most likely.)

(iv) p(spinner lands on an even number) = p(2 or 4) = 0.1 + 0.15 = 0.25

(c) p(spinner lands on 3 and the coin shows Heads) = $0.25 \times \frac{1}{2} = 0.125$

Chapter 16 Histograms

Answer

1

Frequency density

Time (*t* hours)

How to solve these questions

Add a column to the table for the frequency density, and complete it. (Remember, frequency density = frequency ÷ class width, so calculate 16 ÷ 0.5, 28 ÷ 0.5, 26 ÷ 1, 12 ÷ 1, 18 ÷ 2.)

Time t (hours)	Frequency	Class width	Frequency density
$0 \le t < 0.5$	16	0.5	$\frac{16}{0.5} = 32$
$0.5 \le t < 1$	28	0.5	$\frac{28}{0.5} = 56$
$1 \le t < 2$	26	1	$\frac{26}{1} = 26$
$2 \le t < 3$	12	1	$\frac{12}{1} = 12$
$3 \le t < 5$	18	2	$\frac{18}{2} = 9$

Use these frequency densities to draw the histogram.

Answer

2 **(a)**

Frequency density

Speed (km/h)

(b) 65 cars were probably speeding.

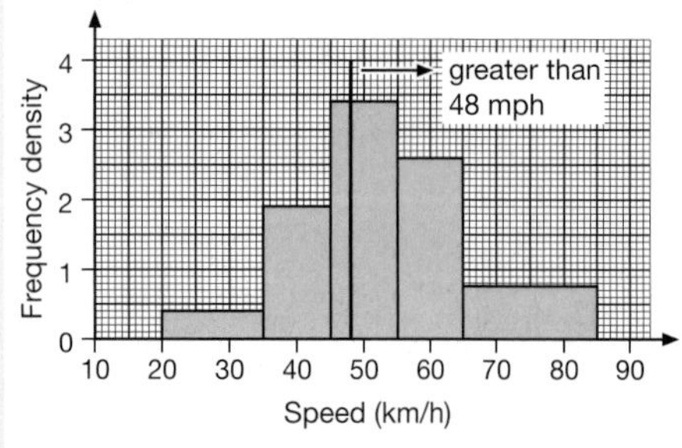

How to solve these questions

(a) Add the frequency densities of 6 ÷ 15, 19 ÷ 10, 34 ÷ 10, 26 ÷ 10, 15 ÷ 20 to the chart. Use the chart to draw a histogram.

Speed (s km/h)	$20 \le s < 35$	$35 \le s < 45$	$45 \le s < 55$	$55 \le s < 65$	$65 \le s < 85$
Frequency	6	19	34	26	15
Class width	15	10	10	10	20
Frequency density	$\frac{6}{15} = 0.4$	$\frac{19}{10} = 1.9$	$\frac{34}{10} = 3.4$	$\frac{26}{10} = 2.6$	$\frac{15}{20} = 0.75$

(b) Draw a line at speed = 48 km/h.

Estimate the area to the right of line and convert it back to give $3.4 \times 7 + 2.6 \times 10 + 0.75 \times 20 = 64.8$

So 65 cars were probably speeding.

Answer

3 See the diagram opposite.

4 **(a)** £453 (to the nearest £)

(b) See the diagram opposite.

How to solve these questions

Draw a table and calculate the respective frequency densities.

Height (cm)	Frequency	Class width	Frequency density
15–19	4	5	$\frac{4}{5} = 0.8$
20–24	6	5	$\frac{6}{5} = 1.2$
25–29	7	5	$\frac{7}{5} = 1.4$
30–39	11	10	$\frac{11}{10} = 1.1$
40–49	9	10	$\frac{9}{10} = 0.9$
50–74	5	25	$\frac{5}{25} = 0.2$
75–99	2	25	$\frac{2}{25} = 0.08$

Remember that the heights are given to the nearest cm so the 15–19 group includes plants from $14\frac{1}{2}$ cm to $19\frac{1}{2}$, so class width = 5, etc.

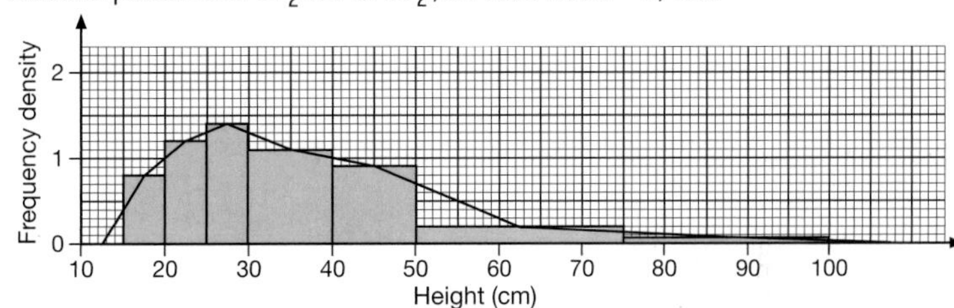

(a) To estimate the mean, draw a table with a column to show mid-interval values (100, 300, 500, 800, 1500) and a column for frequency × price (6×100, 21×300, 17×500, 7×800, 2×1500).

Price (£p)	Mid-interval value	Frequency	MIV × frequency
$0 \le p < 200$	100	6	600
$200 \le p < 400$	300	21	6300
$400 \le p < 600$	500	17	8500
$600 \le p < 1000$	800	7	5600
$1000 \le p < 2000$	1500	2	3000
		$\sum f = 53$	$\sum fx = 24\,000$

$$\text{Mean} = \frac{24\,000}{53} = £452.830\,188\,7 = £453 \text{ (to the nearest pound)}$$

(b) Add columns for the class width and frequency density, like this.

Price (£p)	Mid-interval value	Frequency	MIV × frequency	Class width	Frequency density
$0 \le p < 200$	100	6	600	200	$\frac{6}{200} = 0.03$
$200 \le p < 400$	300	21	6300	200	$\frac{21}{200} = 0.105$
$400 \le p < 600$	500	17	8500	200	$\frac{17}{200} = 0.085$
$600 \le p < 1000$	800	7	5600	400	$\frac{7}{400} = 0.0175$
$1000 \le p < 2000$	1500	2	3000	1000	$\frac{2}{1000} = 0.002$
		$\sum f = 53$	$\sum fx = 24\,000$		

Draw the histogram, using the information in the table.

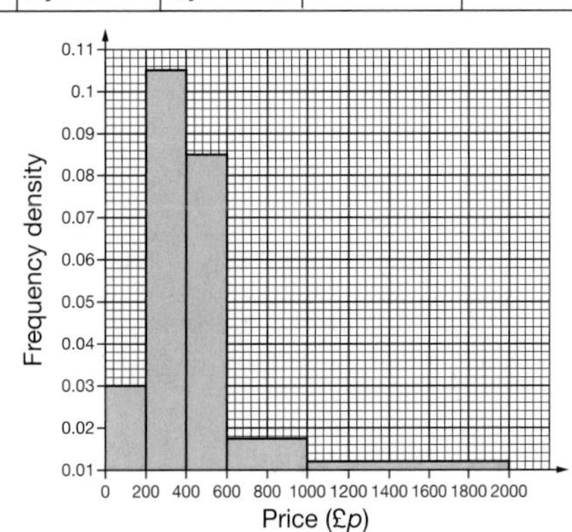

Answer

(c) 24

(d) The mean price of cameras in May 1997 was greater than the mean price of cameras in December 1997.

There was a greater proportion of cameras in the £600–1000 price range in May 1997 than in December 1997.

How to solve these questions

(c) The number of cameras available in May 1997

$= 200 \times 0.005 + 200 \times 0.015 + 200 \times 0.035 + 400 \times 0.0275 + 1000 \times 0.002 = 24$

(d) You are asked to compare, in two ways, the prices of cameras in December 1997 with the prices in May 1997, so you need to look carefully at the information and the graphs and draw two different conclusions from the data presented.

Work out an estimate of the mean price of cameras in May 1997 as you did in part (a).

Mean $= \frac{16300}{24} = £679.1\dot{6}$

5 (a)

Lifetime, x (1000s hours)	Number of bulbs, f	Frequency density
$6 \leq x < 10$	4	1
$10 \leq x < 12$	15	7.5
$12 \leq x < 14$	19	9.5
$14 \leq x < 16$	8	4
$16 \leq x < 20$	4	1

See the diagram opposite.

(a) Complete the frequency density column in the table ($4 \div 4$, $15 \div 2$, $19 \div 2$, $8 \div 2$, $4 \div 4$) and draw the histogram.

Lifetime, x (1000s hours)	Number of bulbs, f	Class width	Frequency density
$6 \leq x < 10$	4	4	$\frac{4}{4} = 1$
$10 \leq x < 12$	15	2	$\frac{15}{2} = 7.5$
$12 \leq x < 14$	19	2	$\frac{19}{2} = 9.5$
$14 \leq x < 16$	8	2	$\frac{8}{2} = 4$
$16 \leq x < 20$	4	4	$\frac{4}{4} = 1$

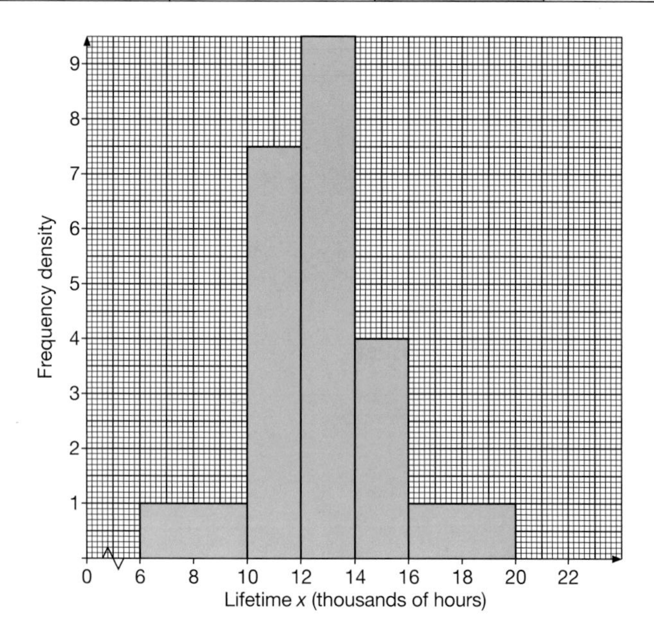

(b) 48%

(b) To find the total number of bulbs, you need to find the total area under the histogram.

$2 \times 3 + 2 \times 6 + 2 \times 8 + 2 \times 5 + 4 \times 1.5 = 50$

The number that last longer than 15 000 hours is:

$1 \times 8 + 2 \times 5 + 4 \times 1.5 = 24$

Proportion $= \frac{24}{50} = 48\%$

(c) The second company's bulbs are more environmentally-friendly.

(c) For the first company, the proportion of bulbs that last longer than 15 000 hours $= \frac{8}{50} = 16\%$.

So the second company manufactures the more environmentally-friendly light bulbs, as the proportion of bulbs that last longer than 15 000 hours is higher.

Chapter 17 Further probability

Answer

1 **(a)**

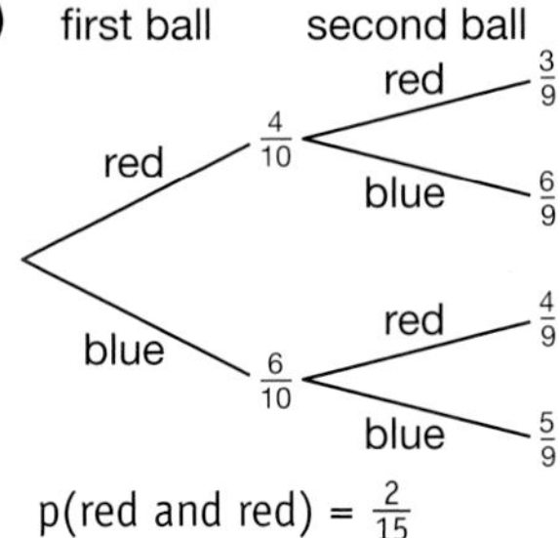

p(red and red) = $\frac{2}{15}$

(b) p(red and blue or blue and red) = $\frac{8}{15}$

2 **(a)** p(red and red and red) = $\frac{1}{22}$

(b) p(RRR) + p(BBB) = $\frac{7}{110}$

(c) p(RRX + RXR + XRR) = $\frac{7}{22}$

3 **(a)** p(Q or R) = 0.55

(b) p(R or S) = 0.45

(c) p(Q or R or S) = 0.75

How to solve these questions

1 **(a)** From the diagram:

probability of obtaining two red balls = p(red and red)

$= \frac{4}{10} \times \frac{3}{9}$

$= \frac{12}{90}$

$= \frac{2}{15}$

(b) From the diagram:

probability of obtaining one ball of each colour

= p(red and blue or blue and red)

$= \frac{4}{10} \times \frac{6}{9} + \frac{6}{10} \times \frac{4}{9}$

$= \frac{24}{90} + \frac{24}{90}$

$= \frac{48}{90}$

$= \frac{8}{15}$

2 **(a)** p(three discs drawn are red) = p (red and red and red)

$= \frac{5}{12} \times \frac{4}{11} \times \frac{3}{10}$

$= \frac{1}{22}$

(b) p(three discs drawn are of the same colour) = p(RRR) + p(BBB)

$= \frac{1}{22} + (\frac{4}{12} \times \frac{3}{11} \times \frac{2}{10})$

$= \frac{1}{22} + \frac{1}{55} = \frac{7}{110}$

(c) exactly two of the discs drawn are red = p(RRX + RXR + XRR)

$= 3 \times (\frac{5}{12} \times \frac{4}{11} \times \frac{7}{10})$

$= \frac{7}{22}$

3 **(a)** p(Q or R) = p(Q) + p(R)

= 30% + $\frac{1}{4}$

= 0.30 + 0.25

= 0.55

(b) p(R or S) = p(R) + p(S)

= $\frac{1}{4}$ + 0.2

= 0.25 + 0.2

= 0.45

(c) p(Q or R or S) = p(Q) + p(R) + p(S)

= 30% + $\frac{1}{4}$ + 0.2

= 0.30 + 0.25 + 0.2

= 0.75

You can give your answers in fractions, decimals or percentages.

Answer / How to solve these questions

4

This question looks tricky, but you can use algebra to help you!

Let p be the probability of a score of 1.

$p(1) = p$
$p(2) = p$
$p(3) = p$
$p(4) = p$
$p(5) = 2p$ (a score of 5 is twice as likely as a score of 4)
$p(6) = 4p$ (a score of 6 is twice as likely as a score of 5)

So $p + p + p + p + 2p + 4p = 1$
$10p = 1$ (total of probabilities of all outcomes = 1)
so $p = \frac{1}{10}$

Now it is quite straightforward.

(a) p(a score of 6) = $\frac{2}{5}$

(a) p(a score of 6) = $4p = \frac{4}{10} = \frac{2}{5}$

(b) p(a score of 5) = $\frac{1}{5}$

(b) p(a score of 5) = $2p = \frac{2}{10} = \frac{1}{5}$

(c) p(a score less than 4) = $\frac{3}{10}$

(c) p(a score less than 4) = p(1) + p(2) + p(3)
$= p + p + p$
$= 3p$
$= \frac{3}{10}$

5 (a)

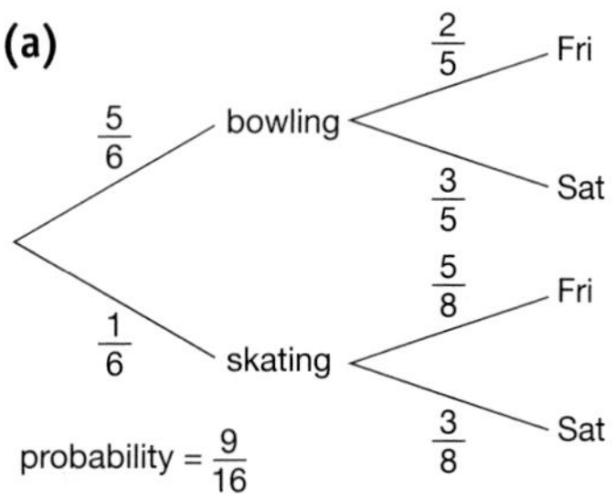

bowling $\frac{5}{6}$: Fri $\frac{2}{5}$, Sat $\frac{3}{5}$
skating $\frac{1}{6}$: Fri $\frac{5}{8}$, Sat $\frac{3}{8}$

probability = $\frac{9}{16}$

(a) Construct the tree diagram and label it.
p(student prefers Saturday) = p(bowling and prefers Sat) or p(skating and prefers Sat)

$= \frac{5}{6} \times \frac{3}{5} + \frac{1}{6} \times \frac{3}{8}$

$= \frac{\cancel{5}^1}{{}_2\cancel{6}} \times \frac{\cancel{3}^1}{{}_1\cancel{5}} + \frac{1}{{}_2\cancel{6}} \times \frac{\cancel{3}^1}{8}$ Try to cancel down if possible.

$= \frac{1}{2} + \frac{1}{16}$

$= \frac{8}{16} + \frac{1}{16}$

$= \frac{9}{16}$

Remember that to add fractions you need a common denominator.

(b) 160 students

(b) $\frac{9}{16}$ of total = 90

$\frac{1}{16}$ of total = $\frac{90}{9}$ = 10 — dividing by 9 to find $\frac{1}{16}$

$\frac{16}{16}$ of total = $16 \times 10 = 160$ — multiplying by 16 to find $\frac{16}{16}$ or whole lot

Total = 160

Chapter 18 Patterns and sequences

Answer / How to solve these questions

1 (a) 1, 3, 5, 7, 9, …

(b) 0, 2, 6, 12, 20, …

(c) $\frac{1}{2}, \frac{2}{3}, \frac{3}{4}, \frac{4}{5}, \frac{5}{6}, \ldots$

Substitute values 1, 2, 3, 4, 5 in the expression for the nth term.

2 (a) 19

(b) (i) $x + 4$

(ii) $x - 4$

(a) You find each term by adding 4 to the previous term.

(b) (i) The next term will be x add 4 or $x + 4$.

(ii) Remember, you can find the term before x by taking 4 away from x, to give $x - 4$.

Answer

3 **(a)** 95, 191

(b) Each term = 2 × previous term + 1

(c) $\frac{n}{n+3}$

4 **(a)** $\frac{5}{11}, \frac{10}{13}, \frac{15}{15}$

(b) 18

5 **(a)** nth term = $3n - 1$

(b) **(i)** $-\frac{1}{2}$, 0

(ii) $n^2 - 2n - 80 = 0$.
$(n + 8)(n - 10) = 0$
$n = -8$ or $n = 10$
so the value of n is 10.

6 **(a)** 10 squares

(b) $2n + 2$

7 The nth pattern has $2n + 1$ matches.

How to solve these questions

3

(a) The sequence is 2, 5, 11, 23, 47, ...

The first differences are +3, +6, +12, +24, ...

Did you recognise that each term = 2 × previous term + 1?

(b) You need to explain how you get from one term of the sequence to the next term.

(c) The numbers on the top form one sequence, the numbers on the bottom form another. Did you spot it?

4

(a) Substitute $n = 1$, $n = 2$ and $n = 3$ in the given formula.

(b) To find the term of the sequence with a value of 2, use the fact that

$$\frac{5n}{2n+9} = 2$$

so $5n = 2(2n + 9) = 4n + 18$

$n = 18$

5

(a) The common difference is +3 so the nth term is $3n - 1$.

(b) **(i)** Substitute 1 and 2 in the formula $\frac{1}{2}(n^2 - 2n)$.

(ii) Solving the quadratic $n^2 - 2n - 80 = 0$ gives $n = -8$ or $n = 10$. Reject the value $n = -8$ as it is unreasonable.

See Chapter 21, *Quadratic equations* for further information.

6

(a)

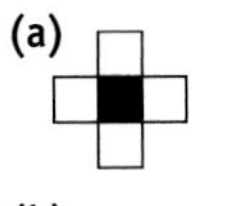
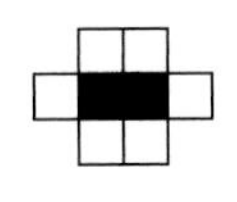
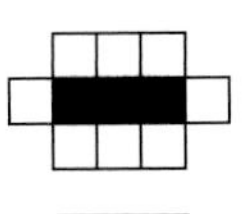
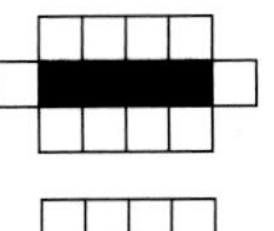

(b)

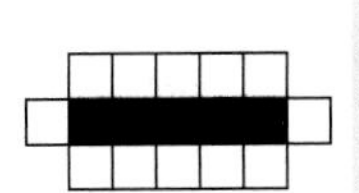

The fourth pattern (pattern 4) gives:
4 + 4 + 1 + 1

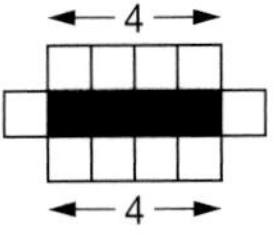

The fifth pattern (pattern 5) gives:
5 + 5 + 1 + 1

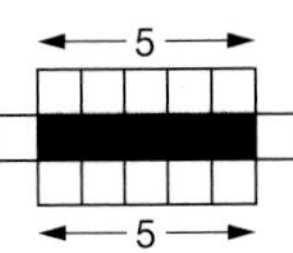

So the nth pattern gives:
$n + n + 1 + 1$

i.e. $n + n + 2$ or $2n + 2$ (simplifying)

7

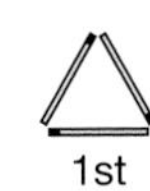
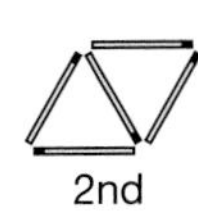
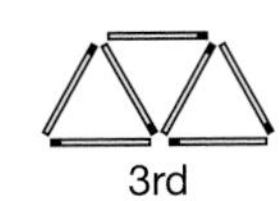
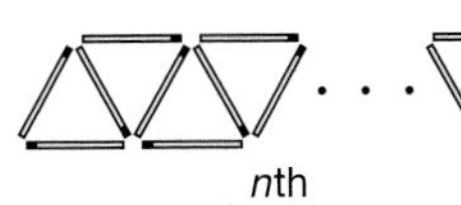

1st	2nd	3rd	nth
Pattern 1	Pattern 2	Pattern 3	Pattern n
3 matches	3 matches + 2	3 matches + 2 + 2	3 matches + 2 + 2 + 2 + ... + 2

There will be $(n - 1)$ lots of 2,
so the nth term $= 3 + (n - 1) \times 2$
$= 3 + 2n - 2$
$= 2n + 1$

Chapter 19 Manipulation of equations

Answer

1

$$t = \frac{p+q}{2}$$
$$2t = p + q$$
$$2t - q = p$$
$$p = 2t - p$$

2 **(a)**
$$r = 4pq - s$$
$$s + r = 4pq$$
$$s = 4pq - r$$

(b)
$$r = 4pq - s$$
$$r + s = 4pq$$
$$4pq = r + s$$
$$q = \frac{r+s}{4p}$$

3 **(a)** $16 \times x + 4 \times (2x + 3)$

$16x + 8x + 12$

$24x + 12$

(b) $24x + 12 = 132$

$24x + 12 - 12 = 132 - 12$

$24x = 120$

$x = \frac{120}{24}$

$x = 5$

4 **(a)** **(i)** $P = 3x + 2 + 2x + 3x + 2 + 2x$

$P = 10x + 4$ cm

(ii) $A = (3x + 2)(2x)$

$A = 2x(3x + 2)$ cm^2

(b) $10x + 4 = 44$ (as $P = 44$)

$10x = 40$

$x = 4$

$A = 2x(3x + 2)$

Substituting $x = 4$:

$A = 2 \times 4(3 \times 4 + 2)$

$A = 112$ cm^2

5 $x = 33$

6 **(a)** £30x

(b) £10x + 10

(c) £40x

(d) $50x + 10 - 30x = 70$

$20x + 10 = 70$

$20x = 60$

$x = 3$

Original cost of share = £3

How to solve these questions

1 Multiply both sides by 2.

Subtract q from both sides.

Rearrange the formula to make p the subject.

2 Remember that you must do the same thing to both sides, to keep the equation correct and balanced.

3 **(a)** 16 small slabs weigh $16 \times x$ kg

4 large slabs weigh $4 \times (2x + 3)$ kg.

Remember to give the answer in its simplest form.

(b) As the total weight of the slabs is 132 kilograms, set it equal to the expression from part (a).

4 Remember that:

the perimeter of a rectangle = $2l + 2b$

the area of a rectangle = $l \times b$.

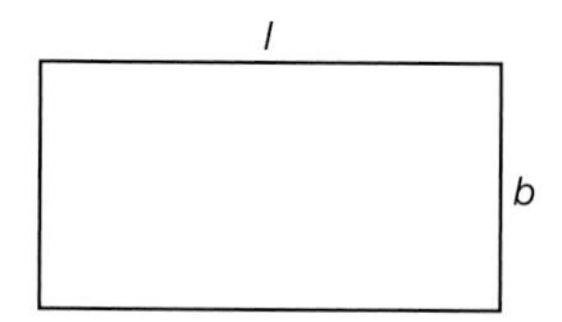

perimeter $= l + b + l + b$

$= 2l + 2b$

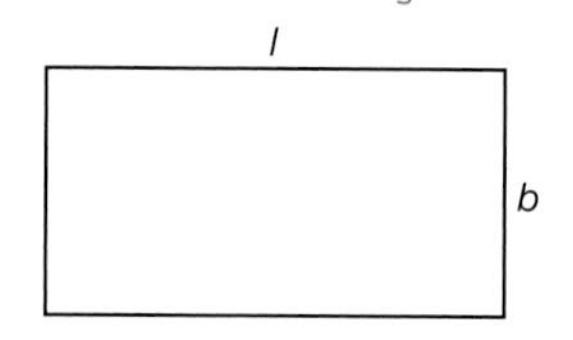

area $= l \times b$

5 $x + 2x + 81 = 180$ The angles of a triangle add up to 180°

$3x + 81 = 180$

$3x = 99$

$x = 33$

6 **(a)** Neil spent $30 \times$ £x.

(b) Originally, each share costs £x.

So if the price of each share has risen by £1:

the new cost is £x + 1.

10 shares cost $10 \times$ £x + 1 or £10x + 10.

(c) Originally, each share costs £x.

So if the price of each share has doubled:

new cost is $2 \times$ £x or £2x.

20 shares cost $20 \times$ £2x or £40x.

(d) Expenditure in pounds = $30x$

Income in pounds = $10x + 10 + 40x = 50x + 10$.

Difference in pounds = 70

So $50x + 10 - 30x = 70$

Solve this equation to find $x = 3$.

Chapter 20 Simultaneous equations

Answer

1

$2x + 3y = 7$
$5x - 3y = 21$
$7x = 28$
$x = 4$

$2x + 3y = 7$
$8 + 3y = 7$
$3y = -1$
$y = -\frac{1}{3}$

How to solve these questions

In this example, the y terms are numerically equal and the signs are opposite. You can add the equations to remove y.

This is the elimination method.

2

$5x + 3y = 19$
$5x - 2y = 4$
$3y - (-2y) = 15$
$5y = 15$
$y = 3$

$5x + 3y = 19$
$5x + 9 = 19$
$5x = 10$
$x = 2$

In this example, the x terms are numerically equal and the signs are the same (i.e. both positive), so you can subtract one equation from the other to eliminate x.

This is the elimination method.

3

$2x + 3y = 1$
$x - 2y = 11$

Substituting $x = 11 + 2y$:

$2x + 3y = 1$
$2(11 + 2y) + 3y = 1$
$22 + 4y + 3y = 1$
$7y = -21$
$y = -3$

$x = 11 + 2y$
$x = 11 + 2 \times -3$
$x = 11 - 6$
$x = 5$

You will probably find that the method of substitution is best for this question.

Here are some possible substitutions:

$x = \frac{1 - 3y}{2}$ (from the first equation)

$y = \frac{1 - 2x}{3}$ (from the first equation)

$x = 11 + 2y$ (from the second equation)

$y = \frac{x - 11}{2}$ (from the second equation)

There is a single term in x in the second equation, so it will be easier to turn this into an expression for x. Then you can substitute for x in the first equation.

4

$x + y = 13$
$x - y = 3$
$2x = 16$
$x = 8$
$x + y = 13$
$8 + y = 13$
$y = 5$

Let your two numbers be x and y. Then form two equations.

- If they have a sum of 13 then $x + y = 13$.
- If they have a difference of 3 then $x - y = 3$.

Solve the equations using the method you like best.

Answer

5

$x + y = 8$

$y = 3x - 4$

Substitute $y = 3x - 4$ in the first equation.

$x + y = 8$

$x + (3x - 4) = 8$

$x + 3x - 4 = 8$

$4x - 4 = 8$

$4x = 12$

$x = 3$

$x + y = 8$

$3 + y = 8$

$y = 5$

The coordinates are (3, 5).

How to solve these questions

You need to solve the two equations simultaneously to find the coordinates of the point of intersection of the lines.

As y is given as the subject in one of the equations, you could choose to use the method of substitution.

Note: You can also solve simultaneous equations by graphical methods, plotting the graph of each of the equations.

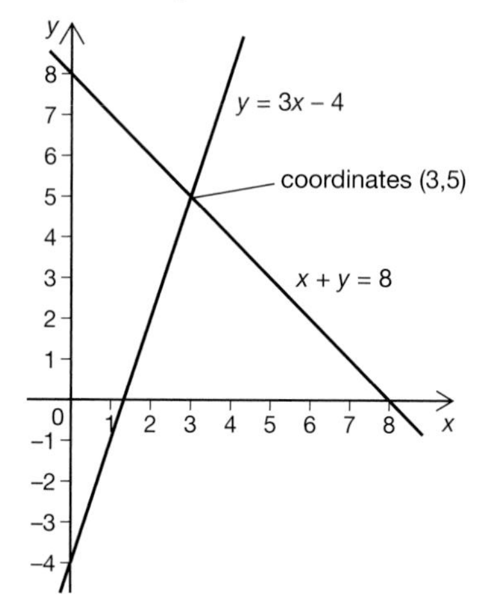

From the graph, the point of intersection is where the equations are solved simultaneously. In this case the point (3, 5) tells us that $x = 3$ and $y = 5$.

Chapter 21 Quadratic equations

Answer

1 **(a)** $x^2 + 9x + 14$
(b) $12x^2 - 29x + 14$
(c) $4x^2 - 4x + 1$

2 **(a)** $(x + 4)(x - 3)$
(b) $(x - 4)(x + 3)$
(c) Does not factorise.

3 **(a)** $x = 4$ or $x = \frac{7}{2}$ (or $3\frac{1}{2}$)
(b) $x = 3$ or $x = 4$
(c) $x = -\frac{1}{2}$ or $x = 3$
(d) $x = -\frac{1}{2}$ or $x = 3$

4 **(a)** $(x - 4)(x - 5)$
(b) $x = 0$ or $x = -2$

How to solve these questions

1 Remember to multiply each term in the first bracket by each term in the second bracket.

(c) Remember that $(2x - 1)^2 = (2x - 1)(2x - 1)$.

2 (c) You can't factorise the quadratic $x^2 + x - 3$.

$x^2 + x - 3 = (x \quad)(x \quad)$

The only numbers that multiply together to give -3 are -3×1 and -1×3.

Neither of these pairs works in the equation.

3 (b) $x^2 - 7x + 12 = 0$ is the same as $(x - 3)(x - 4) = 0$.

(c) $2x^2 - 5x - 3 = 0$ is the same as $(2x + 1)(x - 3) = 0$.

(d) $2x^2 - 5x = 3$ can be written as $2x^2 - 5x - 3 = 0$ which is the same as part (c).

4 (b) You can write $x^2 + 2x = 0$ as $x(x + 2) = 0$.

Answer

How to solve these questions

5 $a = 5$ and $b = -2$

Multiply out $(3x - 1)(x + 2)$.

$(3x - 1)(x + 2) = 3x^2 + 5x - 2$.

Now compare with $3x^2 + ax + b$.

$a = 5$ and $b = -2$.

6 **(a)** $x^2 + 2xy + y^2$

(a) $(x + y)^2 = (x + y)(x + y)$

$= (x + y)(x + y)$

$= x^2 + xy + xy + y^2$ as $yx = xy$

$= x^2 + 2xy + y^2$

(b) 25

(b) Substituting $x = 3.63$ and $y = 1.37$ in (a) above

$3.63^2 + 2 \times 3.63 \times 1.37 + 1.37^2 = (3.63 + 1.37)^2$

$= 5^2 = 25$

7 $(a + b)(a - b) = a^2 - b^2$

$1999^2 - 1998^2$

$= (1999 + 1998)(1999 - 1998)$

$= 3997 \times 1$

$= 3997$

Remember that:

$a^2 - b^2 = (a + b)(a - b)$ and

$(a + b)(a - b) = a^2 - b^2$

This is called the **difference of two squares**.

8 $32x$

You can use the difference of two squares for this problem, too.

$a^2 - b^2 = (a + b)(a - b)$

so that:

$(x + 8)^2 - (x - 8)^2 = \{(x + 8) + (x - 8)\}\{(x + 8) - (x - 8)\}$

$= \{2x\}\{16\}$

$= 32x$

9 **(a)** Area of square $= x \times x = x^2$

Area of triangle $= \frac{1}{2} \times x \times 4$

$= 2x$

Total area $= x^2 + 2x$

The area of the shape is $48\,\text{cm}^2$ so:

$x^2 + 2x = 48$

(a)

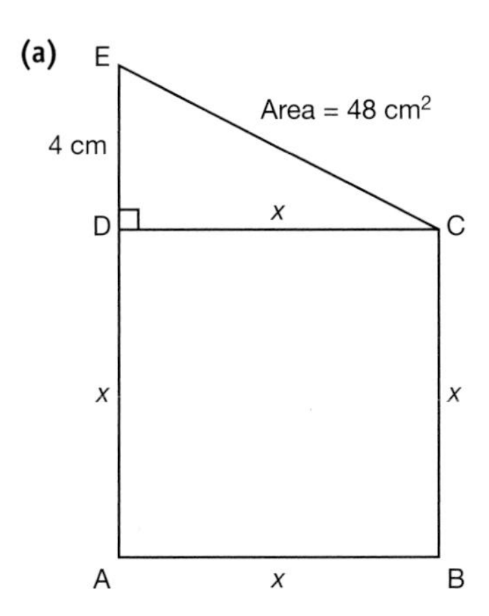

Area of square $= x \times x = x^2$

Area of triangle $= \frac{1}{2} \times x \times 4 = 2x$

Total area $= x^2 + 2x$

Since the area of the shape is $48\,\text{cm}^2$

$x^2 + 2x = 48$

(b) $x^2 + 2x = 48$

$x^2 + 2x - 48 = 0$

$(x - 6)(x + 8) = 0$

so $x = 6$ or $x = -8$

The length must be positive, so length $= 6\,\text{cm}$.

(b) Remember that $x^2 + 2x = 48$ is the same as $x^2 + 2x - 48 = 0$.
Solve the quadratic to find $x = 6$ or $x = -8$.
Ignore $x = -8$ as you can't have a negative length.

Chapter 22 Further manipulation of equations

Answer / How to solve these questions

1 $y = \frac{qx}{x + q}$

Use the flow chart.

Clear fractions by multiplying both sides by the denominator. → $q(x - y) = xy$

Multiply out the brackets and rearrange so all the terms that include the subject are on one side. → $qx - qy = xy$, $qx = xy + qy$, $xy + qy = qx$

Factorise terms containing the subject and divide both sides by the factor. → $y(x + q) = qx$, $y = \frac{qx}{x + q}$

2 **(a)** $\frac{4}{5}$

(a) $s = \frac{1 - (\frac{1}{3})^2}{1 + (\frac{1}{3})^2} = \frac{\frac{8}{9}}{\frac{10}{9}}$

$= \frac{8}{10} = \frac{4}{5}$

(b) $t = \sqrt{\frac{1 - s}{1 + s}}$

(b) Use the flow chart.

Take out the fractions by multiplying both sides by the denominator. → $s(1 + t^2) = 1 - t^2$

Multiply out the brackets and rearrange so all the terms that include the subject are on one side. → $s + st^2 = 1 - t^2$, $st^2 + t^2 = 1 - s$

Factorise terms containing the subject and divide both sides by the factor. → $t^2(s + 1) = 1 - s$, $t^2 = \frac{1 - s}{1 + s}$, $t = \sqrt{\frac{1 - s}{1 + s}}$

3 $s = \frac{p^2t}{p^2 - 1}$

$p = \sqrt{\frac{s}{s - t}}$

$p^2 = \frac{s}{s - t}$ (squaring both sides)

Now rearrange the equation, take out s and find the new expression.

4 $\frac{3x^2 + 15x}{2x^2 + 9x - 5}$

$= \frac{3x(x + 5)^1}{(2x - 1)(x + 5)^1}$

$= \frac{3x}{2x - 1}$

You can always answer these questions in exactly the same way, by factorising (where possible) the numerator and denominator and then cancelling.

5 $(x + a)^2 + b = x^2 + 2ax + a^2 + b$

Compare with

$x^2 - 6x + 13$

$a = -3$

$b = 4$

Work out the expression on the left-hand side and compare coefficients of x^2, x and the constant.

Coefficient of x: $2a = -6$

$a = -3$

Constant: $a^2 + b = 13$

$(-3)^2 + b = 13$

$9 + b = 13$

$b = 4$

Answer

6 $\frac{1}{x-2} + \frac{1}{x+3}$

$= \frac{1(x+3)}{(x-2)(x+3)} + \frac{1(x-2)}{(x-2)(x+3)}$

$= \frac{(x+3)+(x-2)}{(x-2)(x+3)}$

$= \frac{2x+1}{(x-2)(x+3)}$

How to solve these questions

You need to start by writing $\frac{1}{x-2} + \frac{1}{x+3}$ as a single fraction.

Set both terms over a common denominator of $(x-2)(x+3)$ by forming equivalent fractions.

$\frac{1}{x-2} = \frac{1(x+3)}{(x-2)(x+3)}$

$\frac{1}{x+3} = \frac{1(x-2)}{(x-2)(x+3)}$

Now you just need to add the expressions and simplify the numerator.

7 $\frac{3 \times 7(2x-3)}{7(x+2)(2x-3)} - \frac{2 \times 7(x+2)}{7(x+2)(2x-3)}$

$= \frac{(x+2)(2x-3)}{7(x+2)(2x-3)}$

First, you need to rewrite each term, setting it over a common denominator of $7(x+2)(2x-3)$.

$\frac{21(2x-3)}{7(x+2)(2x-3)} - \frac{14(x+2)}{7(x+2)(2x-3)}$

$= \frac{(x+2)(2x-3)}{7(x+2)(2x-3)}$

$\frac{42x-63}{7(x+2)(2x-3)} - \frac{14x+28}{7(x+2)(2x-3)}$

$= \frac{2x^2+x-6}{7(x+2)(2x-3)}$

Multiply out the brackets.

$(42x-63) - (14x+28) = 2x^2 + x - 6$

$2x^2 - 27x + 85 = 0$

$(2x-17)(x-5) = 0$

so $x = \frac{17}{2}$ or $x = 5$

Then multiply throughout by $7(x+2)(2x-3)$ to remove the fraction.

Use brackets to keep terms together, before simplifying.

Simplify the resulting expression. You should find it reduces to a quadratic.

Factorise the quadratic.

Solve to find the value of x.

Chapter 23 Further quadratic equations

Answer

1 $(2x-1)(x+4) = 0$

$x = \frac{1}{2}$ or $x = -4$

How to solve these questions

You can rewrite $2x^2 + 7x = 4$ as $2x^2 + 7x - 4 = 0$.

Then you can factorise it, in the usual way.

2 $x = 0$ or $x = \frac{2}{3}$

You can factorise the equation $3x^2 - 2x = 0$ as $x(3x-2) = 0$.

You will probably find that this is easier than trying to apply the formula.

3 $\frac{2}{x+2} = \frac{x-1}{x}$

$\frac{2x}{x+2} = x - 1$

$2x = (x-1)(x+2)$

$2x = x^2 + x - 2$

$x^2 - x - 2 = 0$

$(x-2)(x+1) = 0$

so $x = 2$ or $x = -1$

You can remove the fractions by multiplying both sides of the equation by x, then by $x + 2$.

Then you can multiply out the brackets.

Collect like terms on the same side, and form a quadratic.

You should be able to factorise the quadratic and solve it.

Note: Read Chapter 22 *Further manipulation of equations*, if you need more practice with algebraic fractions.

Answer

How to solve these questions

4 $x^2 + 4x - 9 = [(x + 2)^2 - 4] - 9$
$= (x + 2)^2 - 13$

As $(x + 2)^2 = x^2 + 4x + 4$
Then $(x + 2)^2 - 4 = x^2 + 4x$ or $x^2 + 4x = (x + 2)^2 - 4$

$x^2 + 4x - 9 = 0$

$\Rightarrow (x + 2)^2 - 13 = 0$ — From above.

$\Rightarrow (x + 2)^2 = 13$ — Add 13 to both sides.

$\Rightarrow (x + 2) = \pm\sqrt{13}$ — Take the square root on both sides and remember both positive and negative roots.

$\Rightarrow x = -2 \pm \sqrt{13}$ — Subtract 2 from both sides.

5 For $x^2 - 2x - 1 = 0$
$a = 1,\ b = -2,\ c = -1$

Substituting in the formula:

$$x = \frac{-b \pm \sqrt{b^2 - 4ac}}{2a}$$

$$x = \frac{-(-2) \pm \sqrt{(-2)^2 - 4\times1\times-1}}{2 \times 1}$$

Substitute in the formula.

$$x = \frac{+2 \pm \sqrt{4 + 4}}{2}$$

Remember $-(-2) = +2$.

$$x = \frac{2 \pm \sqrt{8}}{2}$$

$$x = \frac{2 + \sqrt{8}}{2} \quad \text{or } x = \frac{2 - \sqrt{8}}{2}$$

This simplifies to $x = 1 + \frac{\sqrt{8}}{2}$ or $x = 1 - \frac{\sqrt{8}}{2}$

or $x = 1 + \sqrt{2}$ or $x = 1 - \sqrt{2}$

as $\frac{\sqrt{8}}{2} = \frac{\sqrt{4 \times 2}}{2} = \frac{2\sqrt{2}}{2} = \sqrt{2}$.

$x = 2.414\,213\,562$
$x = 2.41$ (2 d.p.)

$x = -0.414\,213\,562$
$x = -0.41$ (2 d.p.)

6 **(a)** $(3x + 4)(x - 2)$
(b) $x = -\frac{4}{3}$ and $x = 2$
(c) $-1, 0, 1, 2$

(a) You should be able to factorise the quadratic quite easily.

(b) Solve the equation by setting each factor equal to 0.

(c) The values $x = -\frac{4}{3}$ and $x = 2$ divide the number line into three distinct sections i.e. $x < -\frac{4}{3}$, $-\frac{4}{3} < x < 2$ and $x > 2$.

You can find the answer by testing integer values in these regions.

If you prefer, you could draw a graph, like this.

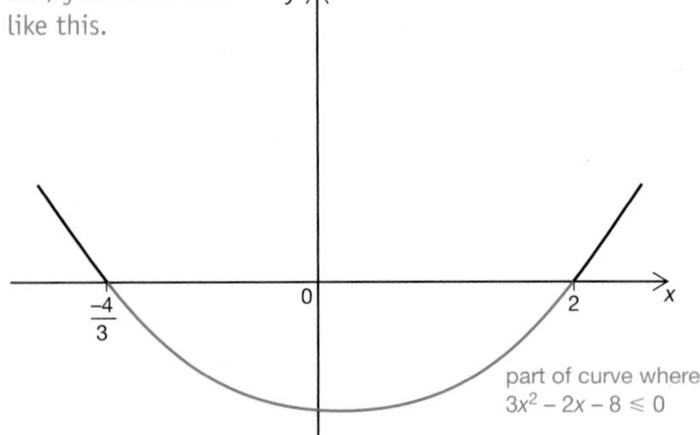

Remember that the question only asks for integer values.